The
Environmental
Impact
Statement
Process

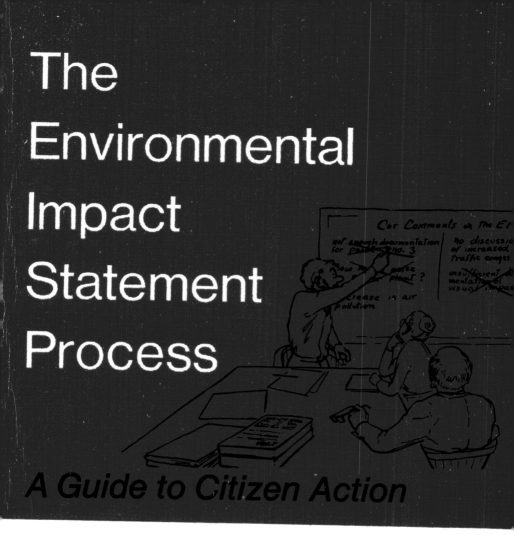

A Guide to Citizen Action

NEIL ORLOFF

INFORMATION RESOURCES PRESS

The Environmental Impact Statement Process

A Guide to Citizen Action

by
NEIL ORLOFF

Illustrations by
DIANE EDWARDS LA VOY

INFORMATION RESOURCES PRESS WASHINGTON, D.C. 1978

Information Resources Press
2100 M Street, N.W.
Washington, D.C. 20037

Library of Congress Catalog Card Number 78-53825

ISBN 0-87815-021-8

Foreword

The National Environmental Policy Act (NEPA) offers major opportunities for the prevention of environmental damage. Through its impact statement process, federal agencies are now required to forecast the environmental consequences of projects and their alternatives, and to make these forecasts available to the public. This has made it possible to anticipate actions that threaten our priceless environmental heritage and also has opened up, for public scrutiny, the choices that are available to our federal agencies. These requirements alone, however, cannot achieve the goal of better government decisions. The past eight years have demonstrated unquestionably the need for citizen involvement if this goal is to be realized.

Citizens are, after all, the only "experts" capable of making the basic value choices that underlie federal programs and projects. It is one thing for an impact statement to identify the environmental consequences of a project; it is another to decide whether they are acceptable. A government official in Washington, D.C. has a different perspective on the economic, environmental, and other trade-offs involved than a resident of the region who will benefit from the proposed project or a citizen in the immediate area affected by it. NEPA makes it possible to achieve a new balance among these perspectives. But, if this is to occur, citizens must make themselves heard.

The Council on Environmental Quality (CEQ) has worked to foster public participation since it first issued impact statement guidelines in 1971. In periodically revising the guidelines, CEQ has expanded the role of citizens and sought to make information more readily available to them. Most recently, in the proposed new regulations to improve the impact statement process, CEQ has emphasized the importance of early and effective citizen involvement and stressed the responsibility of agencies to facilitate this involvement. Opportunities for citizens to use impact statements promise to steadily increase.

This excellent book gives citizens the information needed to make their participation in the impact statement process truly effective. It lays out clearly and accurately the particular points and techniques open to citizens to change federal plans and decisions. For this guidance, Professor Orloff, a recognized expert in the impact statement process, is to be congratulated. His writing reflects the unique perspective of a career which has combined government service as an administrator and lawyer, academic research and teaching, and legal representation of environmental interests for both citizen groups and businesses. He has served the public well through his explanation of how individuals concerned about environmental degradation can help implement the goals of the National Environmental Policy Act.

April 10, 1978   Charles Warren
Washington, D.C.   CHAIRMAN
                                                     President's Council on
                                                     Environmental Quality

Preface

Responding to the public outcry during the late 1960s for increased protection of the environment, Congress passed several far-reaching and highly innovative environmental laws. The earliest and most popular piece of legislation to emerge as a result of this heightened public concern was the National Environmental Policy Act (NEPA). Signed into law on January 1, 1970, NEPA established the Council on Environmental Quality (CEQ) to advise the president on environmental issues and to review federal programs in terms of the country's environmental policies. It also created the environmental impact statement process, which requires every federal agency to report on any major action it proposes—whether it involves funding a local highway project, licensing a new power plant, or constructing a federal building—that would significantly affect the environment.

The basic idea underlying the impact statement process is simple: the Federal Government should understand how a project will affect the environment *before* it is approved. The process requires that agencies "look before they leap." In addition, since decisions on what is an acceptable amount of environmental damage require very difficult value judgments, the process requires that agencies consult with members of the public. Citizens are given an opportunity to review the environmental report prepared by a federal agency and to indicate, through comments on the report, the acceptability of the environmental damage. The process encourages much greater citizen and local government involvement in federal decision making than was previously possible.

v

It is difficult to criticize these objectives. It is also difficult to fault those initially charged with implementing this new environmental program. Most of them were young, intelligent, and hardworking, and they tended to abhor bureaucratic red tape. Their concern was to ensure that the thrust of the legislation was realized without the creation of a myriad of rules or superfluous paperwork.

While NEPA's beginning was auspicious, anyone closely connected with the impact statement process can attest that it has evolved into something much different than this start might suggest. The impact statement process has become a complex government program; the federal regulations currently are hundreds of pages long. Indeed, the field has become so complex that a new profession has been spawned by the act. Today, there are environmental consultants and environmental lawyers who specialize in NEPA and the impact statement process.

This book has two major objectives. First, it is aimed at clearly presenting how, in practice, the environmental impact statement process operates. It has been written for the local government official, the staff member of a federal agency charged with writing an environmental impact statement, the beleaguered state official trying to cope with NEPA's requirements, the citizen group trying to stop a project, or a lawyer trying to work through the differences between what the law says and how it operates. It explains how a major federal environmental program functions.

The second objective is to show citizens how they can use the environmental impact statement process to influence government decision making. The book is a guide to action. It describes not only what the requirements are but also what factors tend to influence agency implementation of the process; it presents many of the forces that act on agency officials during their review of a project.

An indication of what the book *is* also requires an indication of what the book *is not*. Agency decision making almost always involves many steps, only one of which is the environmental impact statement process. Decisions on projects such as highways, power plants, and housing developments occur in their own, much larger, frameworks, which are not discussed in this book. While the book shows how the impact statement process is tied into them, it is up to the reader to investigate what other steps will be taken in connection with the specific project with which he is concerned. This book also does not focus on techniques for citizen action. While suggestions are presented for how citizens can use the impact statement process to influence government decision making, it is beyond the scope of this book to present, in great detail, how to mount a campaign against government agencies. Factors such as organizing, the use of media, the cost of litigation, fundraising, mapping out strategies, and similar matters are touched on only briefly. Appendix F provides suggestions for further readings in this area.

NEPA has given rise to a number of sharply different perspectives on what effect the statute has had. Some people have called it solely a paperwork exercise. Some have looked at the requirements for impact statements primarily as potential legal pitfalls to be used by lawyers in their litigation to stop a project. Some have suggested that NEPA is a Trojan horse—that it has diverted the attention of environmentalists from the "real issues" surrounding projects. Some have viewed the statute as eroding the ability of states and business groups to control the development of their projects and transferring much of this power to environmental groups. All these views have some validity; however, as with the blind men and the elephant, none of them are by themselves an accurate reflection of the overall effect of the statute.

My experiences with NEPA and the environmental impact statement process have been in a number of different capacities. I have worked as a government official on the basic implementation of the statute—helping to establish and administer the Environmental Protection Agency's program for preparing and reviewing statements and drafting major parts of the impact statement guidelines issued in 1973 by the Council on Environmental Quality. I have advised businesses, state and local governments, and citizen groups on the use of the statute and have conducted litigation in the impact statement area. I also have approached the process, while a professor of environmental law, through the eyes of a teacher and researcher. My experiences with NEPA over the last seven years lead me to conclude that the process is much more complicated than any single one of these views suggests.

Two of the widespread, but I think erroneous, views require comment. The first is that NEPA is solely a paperwork exercise. According to this view, impact statements are public-relations devices issued to comply with federal regulations; they rationalize rather than assist agency decision making. My opportunity to observe the process from its inception leads me to the conclusion that, while this sometimes is certainly true, it is not always the case; and, it is definitely becoming less so as agencies become more familiar with the law *and* as successful citizen challenges to insufficiently considered proposals create pressure for a meaningful rather than superficial weighing of environmental hazards. This view ignores the major changes that have been undertaken within the Federal Government in response to the statute. Agencies such as the Forest Service and the Nuclear Regulatory Commission have substantially changed their programs as a result of NEPA. They have modified both their planning activities and the manner in which they appraise individual projects. Moreover, throughout the Federal Government, many projects never surface today because of the act. Their sponsors are aware that they are unlikely to survive the public scrutiny that the impact statement process demands. Two years ago, in its study on the implementation of NEPA by 70 federal agencies, the Council on Environmental Quality reported on a number of agency decisions and practices which have changed as a result of the statute; and, government officials repeatedly have testified before Congress on the substantive effect of the statute on their programs. While the impact statement process is sometimes a paperwork exercise, there is abundant evidence of significant changes that are beginning to occur as a result of the act.

The second view that requires comment is that NEPA is primarily beneficial to environmentalists as a basis for litigation. As the book shows, there are many ways to influence government decision making through the environmental impact statement process without resorting to litigation. These other, earlier modes of intervention are often much more effective in eliciting favorable public opinion and agency response. Moreover, there are two major reasons for avoiding litigation. First, lawsuits usually are very expensive. A group should normally not consider embarking on legal action unless it has accumulated at least several thousand dollars, and most major lawsuits under NEPA run between $25,000 to $100,000. Secondly, while "successful" lawsuits under NEPA have delayed projects, they have not permanently halted them. Litigation under NEPA does serve to increase the pressure on an agency, and delay is sometimes tantamount to termination. There are other ways, however, to delay projects. When litigation is used solely as a delaying tactic, frequently it only temporarily halts the project; other tools must then be used to ultimately resolve the controversy.

Many people assisted in the preparation of this book. The members of Cornell's Project on Environmental Impact Statements helped give the book its early shape. Gene Duvernoy wrote a draft of Chapter 3; Jon Bart wrote a draft of Chapter 4; Wayne Marks wrote Appendix G; and John Stanturf and Amy Freirich helped with the research. Carin Rundle cheerfully typed one revision after another.

Diane Edwards La Voy developed the illustrations and continually asked the simple but penetrating questions. Mike Elsass helped transform much of the rough text into more clearly presentable ideas. Gene Allen and Nancy Moran at Information Resources Press applied the finishing touch to the entire manuscript before its publication.

Malcolm Baldwin, Raymond Bowers, Ernest Callenbach, Ernest Gellhorn, David Kaye, and Jane Magee reviewed a preliminary version of the manuscript and provided a number of helpful comments. The Rockefeller Foundation and the Office of Water Research and Technology at the Department of the Interior provided the funding for the Cornell Project on Environmental Impact Statements. Without them, and the encouragement and support of Ralph Richardson and Gary Toenniessen at the Rockefeller Foundation, Cornell's impact statement project would not have been started, and this book would not have been written.

My wife, Jan, helped with the conceptualization of the book and the rewriting of a substantial part of the text. She also provided a great deal of encouragement throughout the development of the book.

While the responsibility for the final version is mine, a large measure of any credit is due to these people. They contributed generously of their time, and they have my thanks and appreciation.

April 5, 1978
Tempe, Arizona

Neil Orloff

Contents

xi

1 Introduction

Finding Out About Projects

Today, many projects that affect entire communities seem to arise overnight. Local citizens learn about a proposed power plant by noticing a small article in the newspaper, or they see a notice on the bulletin board at the post office announcing a new sewage-treatment plant to be constructed in their county. A friend may casually mention at lunch that he has heard that there are plans to expand a nearby highway, or the project may suddenly become the subject of heated discussion at a party. People frequently find out about new projects by chance, or when they see bulldozers beginning to change the landscape.

If a citizen does find out about a new project before construction starts, he may be concerned with where it will be located, how big it will be, or how much smoke, odor, noise, or water pollution will result. Unfortunately, the citizen trying to delve further into the matter is likely to become frustrated. In the planning stages of a large project, usually little information is available to those most likely to be affected—the members of the local community. Even if a few facts are available, a balanced perspective on the situation may be impossible to achieve.

Moreover, a citizen's instinctive apprehension upon learning of a project may be well founded. Major new projects often are heralded as only benefitting the community. The artist's sketch of a proposed project is likely to show a sleek highway or a well-landscaped

1

On paper, most projects—power plants, highways, dock improvements—look benign. But when built, some can have serious effects on the environment.

power plant that is almost entirely hidden from view by trees. If questions are raised, the need for a fast-transportation corridor or for additional power is likely to be mentioned, as well as the new jobs and taxes that the project will bring to the community. Yet the citizen may be concerned that the project might not turn out entirely as it is presented: It might produce significant side effects, such as air, water, or noise pollution; it might displace people from their homes; or it might take away part of a park or wildlife refuge. On paper, most major projects look benign. But after they are completed, they could have serious effects on the environment.

As the environmental consequences of major projects have become increasingly clear, a new mood has arisen around the country. Many people want to have a say in matters concerning the development of their community. They feel strongly about the alternative uses to which the nearby rivers and lakes, farms, parks, and wildlife refuges, or just the vacant lot down the street, might be put.

Today, many people have ideas about the kind of development that should occur. They want to have a say in what happens, but may not know how to begin.

They have ideas about the kind of development that should—or should not—take place. Often, however, they don't know how to make their views known, particularly if they have not previously participated in government decision making. They want to preserve their local environment, but they just don't know how to go about it or even where to begin.

Three Tasks

Three major tasks confront any citizen who wants to influence the development of his community. They are substantial steps that must be taken if the citizen is to play an effective role in the decision-making process.

1. The first task is to obtain all pertinent information concerning a proposed project. This is difficult because information may be scattered among different government agencies, contractors, and consulting firms. In fact, it may not be immediately clear just who is involved in the project and who is ultimately responsible for it.

2. The second task is to analyze the information collected and to make an independent judgment concerning the project. The citizen must decide whether, on balance, it appears desirable for the community. This can be more difficult than it may seem, because much of the information available will be technical, and it may be incomplete or one-sided.

3. The third task is to find out who will make the final decisions on the project and to make one's views known to those people— who may be scattered among different organizations, and who may be 3,000 miles away.

The average citizen facing all of these obstacles may decide to just leave it to the "experts." Scientists, planners, and engineers will, after all, have insights into technical matters that most citizens cannot equal; however, decisions concerning large projects such as highways, airports, and power plants inevitably involve value judgments and trade-offs on what is best for a community. Experts certainly are no more capable of making *these* decisions than the local citizens who will be directly affected. Further, because of NEPA, established procedures now exist that citizens can use to learn about and comment on certain types of projects.

THREE DIFFICULT TASKS

*Finding out all about
the proposed project. . .*

*Studying the information,
and deciding whether to support
or oppose the project,
and what changes or further study
to recommend. . .*

*Expressing those views
to the right people,
at the right time,
and in the right way.*

The National Environmental Policy Act

Prior to January 1, 1970, the Federal Government provided little support to citizens concerned about proposed federal projects. Except in isolated cases, it did not require that the environmental effects of a proposed government action be studied, that information on the project be made available to the public, or that the public's views be solicited.

On that date, however, the National Environmental Policy Act (see Appendix A) was signed into law. The stated purpose of the act is to

. . . encourage productive and enjoyable harmony between man and his environment; to promote efforts which will prevent or eliminate damage to the environment and biosphere and stimulate the health and welfare of man; to enrich the understanding of the ecological systems and natural resources important to the nation . . . (Sec. 2)

There are three major provisions in the act:

1. It declares that each person has a right to a healthful environment, and that it is in part the responsibility of the Federal Government to ensure that the environment is protected.

2. It establishes the Council on Environmental Quality, a small agency in the Executive Office of the President, whose head is the president's advisor on the environment. The council's responsibility is "to formulate and recommend national policies to promote the improvement of the quality of the environment." Its duties include gathering information, reviewing programs, conducting investigations, and assisting the president in the preparation of an environmental quality report to be issued annually to the Congress.

3. It establishes the environmental impact statement process. Every federal agency is now required to "include in every recommendation or report on proposals for legislation and other major federal actions significantly affecting the quality of the human environment a detailed statement" which includes a description of the environmental impact of the proposed action, unavoidable adverse effects which would result should the action take place, possible alternatives, and a discussion of short-term versus long-term advantages of the proposal. The agency must then circulate the analysis publicly and consider both the analysis and public

responses to it when making its final decision on the proposal. This analysis is called an environmental impact statement or EIS.

The development of the impact statement process represents a major effort by the Federal Government to assemble information on the environmental consequences of proposed actions and to involve the public in decision making. More than 1,000 government officials are involved in administering the process, and federal agencies spend nearly $200 million per year to prepare and review environmental impact statements. Since 1970, more than 9,000 impact statements have been written, covering actions as diverse as funding a highway, building a housing project, or licensing a power plant. Approximately 100 statements are now being written every month, and they range in length from twenty-five to thousands of pages, depending on the complexity of the project, the environmental problems that are likely to arise, and the conscientiousness of the agency. The impact statement is, potentially, one of the most important tools citizens can use to protect both their communities and natural resources from thoughtless or unnecessary abuse.

The Environmental Impact Statement Process

An environmental impact statement is, simply, a prediction of the potential environmental consequences of a proposed federal action, together with a discussion of possible alternatives to the action. Although the idea may be simple, the process involved in writing, reviewing, and acting on the statement can become quite complex. The process involves four major stages:

Stage 1. First, the federal agency must decide whether a statement must be prepared for the proposed project. The law requires an impact statement whenever an agency proposes to take an action that significantly affects the environment. This leaves great latitude for interpretation, but each agency does have specific guidelines.

Stage 2. If an impact statement is required, the second stage involves the preparation of a draft statement—a preliminary report on the environmental consequences associated with the project. If a statement is not required, the agency can proceed with the project; although, if the decision not to prepare a statement is questionable, the agency will frequently document the basis for its determination.

Stage 3. The agency sends the draft statement to all groups having an interest in the proposed action: other federal agencies, including CEQ; state and local governments; local business and industry; private citizens; and interest groups, such as the Chamber of Commerce or the Sierra Club. These parties are allowed a reasonable amount of time, usually 45 days, to study the draft and to send their comments to the agency. The agency then prepares the final impact statement by revising the draft to incorporate the agency's response to the comments received. In most cases, the comments themselves also are included in the final statement.

Stage 4. The agency now proceeds to make a decision on the project, based on the analysis contained in the impact statement as well as the nonenvironmental factors. NEPA requires that the final

In writing an environmental impact statement (EIS), federal agency personnel draw on the results of many kinds of research in order to predict the environmental effects of a proposed action and its alternatives.

statement accompany the proposed action throughout the remainder of the agency's decision-making process.

Relationship of NEPA to Projects of State and Local Governments and the Private Sector

Only federal agencies are required by NEPA to write environmental impact statements on their projects; state and local agencies and private businesses are not compelled by law to write statements. NEPA, however, does apply to those projects in the private sector and those initiated by state and local government agencies in which the Federal Government is involved, which explains why impact statements often are written on such projects as privately owned power plants or state highways.

For the purposes of NEPA, there are two major types of federal involvement in otherwise nonfederal actions or projects:

1. If any federal funding is involved in the project, the project is considered to be federal for the purposes of NEPA, and an impact statement must be written if the project will seriously affect the environment. Highways, sewage-treatment plants, recreational facilities, and housing projects are a few examples of projects for which federal funds are used.

2. If the project requires a federal license or permit, the project is considered to be federal for the purposes of NEPA, and an impact statement must be written if the environment will be significantly affected. Offshore drilling for oil, building a power plant, and harvesting timber in a national forest all require federal permits.

It is important to keep in mind that, although a project may not *seem* to have any connection with the Federal Government, it may be covered by NEPA. Conversely, if a project has no connection with the Federal Government, it will not be affected by NEPA, no matter how seriously it may damage the environment.

Unfortunately, it is not always easy to determine whether an action is "federal" under the rules of NEPA. In most cases where funding is involved, a few phone calls to federal agencies should turn up the necessary information. But, in the case of permits or licenses, even those individuals involved in the project may not know whether it is necessary to apply for one, so some fairly extensive research may be required to determine "federal" involvement.

A project becomes a "federal action" when either funding or permits are requested from the Federal Government.

State Requirements

Twenty-nine states and Puerto Rico have adopted, by statute, executive order, or administrative regulation, an environmental impact statement process similar to NEPA. Although a few of the states do not refer to their requirements as an "impact statement process," these states do follow the general NEPA pattern, through which the public is involved in the review of projects that significantly affect the environment.

These "little NEPAs" apply to activities in which a state agency is involved, such as expanding a highway, changing the classification of a stream, or approving the use of a site as a solid-waste landfill; some also apply to activities in which a local agency is involved, such as issuing a building permit or amending a zoning ordinance. Therefore, even if an activity does not involve the Federal Government, it may—because of state requirements—still have to go through an environmental impact statement process.

As of April 1, 1978, the states with "little NEPAs" were Arizona, Arkansas, California, Connecticut, Delaware, Florida, Georgia, Hawaii, Indiana, Maine, Maryland, Massachusetts, Michigan, Minnesota, Mississippi, Montana, Nebraska, Nevada, New Hampshire, New Jersey, New York, North Carolina, North Dakota, South Carolina, South Dakota, Utah, Virginia, Washington, and Wisconsin. Appendix G contains, for each of these states, a reference to the statute, executive order, or administrative regulation establishing the impact statement process, and the name, address, and phone number of an individual who can be contacted for information on the state's "little NEPA."

While the impact statement process in each of these states is similar to the one established by NEPA, all of them differ in some respects from the federal statute as well as from each other. For example, the California process applies to projects of all state and local agencies, whereas the Arizona process applies only to projects of the Game and Fish Commission and the Power Plant Transmission Line Siting Committee. The Nebraska process applies only to projects of the Department of Roads. The New York process requires agencies to include an analysis of growth-inducing effects in their statements. The Connecticut process requires agencies to include a formal cost-benefit analysis in their statements. Although this book generally applies to all activities that are subject to an environmental impact statement process, it has been written specifically with reference to NEPA. A person involved with a project under a state's "little NEPA" should check the particular requirements of that state.

The Citizen's Role

The obstacles a private citizen must overcome in order to influence a government decision—namely, getting information about the proposed project, deciding whether to support or oppose it, and making his views known to the right people—are now much easier to overcome, due to NEPA and its impact statement process. By law, the government must analyze the environmental consequences of proposed actions and release this information to the public. While the statement is unlikely to tell the concerned citizen whether to be for or against the action, it will provide much of the necessary information on which to base such a decision. Finally, through the impact statement process, citizens have the opportunity to voice their opinions, which must then (by law) be included in the final statement that will be used by government officials when making their decision on the action.

Two Examples of NEPA in Action

Here are two examples of how citizens have used the impact statement process:

The Southern Crossing. In the late 1960s, the California State Division of Bay Toll Crossings announced plans to build a vehicle toll bridge across San Francisco Bay, five miles south of the San Francisco-Oakland Bay bridge. Because no federal funding was involved, state and local officials assumed that the project was not federal for purposes of NEPA. But then a Department of Transportation employee discovered that an application for a Coast Guard permit had been filed and that the bridge could not be built until the permit was granted. While the Coast Guard was skeptical that issuance of the bridge permit was a "major federal action significantly affecting the environment," it was persuaded that the permit might be so construed; therefore, an impact statement was prepared.

The draft statement indicated that no direct environmental effects were anticipated as a result of the bridge construction, but that there could be serious secondary effects. Construction of additional highways to provide access to the bridge might damage the shoreline ecology, and construction of the bridge itself could lessen use of the Bay Area Rapid Transit System, which was not yet fully operational.

A number of Bay Area citizen groups already were opposed to the project because of the upheaval it would cause in neighborhoods in the proposed access areas. The draft statement provided these groups with a strong weapon to fight the proposal. Based on information contained in the draft statement, newspaper articles

CITIZENS' ROLES IN THE
EIS PROCESS

Federal agencies prepare a draft environmental impact statement

. . .which is sent to all groups with an interest in the proposed action: local governments, other federal agencies, business ps, environmentalists, and so on.

Citizens comment on the draft.

These comments, and the agency's response, must be included in the final statement which government officials must consider when they decide about the project.

and editorials were written, and the Council on Environmental Quality asked the Department of Transportation to delay the granting of the permit pending further discussion. The San Mateo Board of Supervisors then reversed an earlier motion in support of the bridge, and the State Assembly voted 53–7 against the project. A statewide election ultimately decided against construction of the bridge.

In this case, public sentiment already was strongly against the bridge. The impact statement, which specifically outlined many of the negative effects of construction, helped citizen groups to document their positions. They used the impact statement to increase public support and to put pressure on the Department of Transportation.

American Cyanamid. In 1956, American Cyanamid completed construction of a new chemical factory on the outskirts of Savannah, Georgia and began dumping sulfuric acid, iron sulfates, and heavy metal wastes into the Savannah River. In the late 1960s, the newly created Georgia Water Quality Control Board (GWQCB) ordered the company to find an alternative method for waste disposal.

In 1970, American Cyanamid proposed barging the wastes far out to sea and dumping them in international waters. In order to do so, the company required docking facilities. Under an 1899 statute written to protect and maintain navigable waters, permission was required from the Army Corps of Engineers for any construction affecting the Savannah River; accordingly, the company filed a permit application.

Shortly before the application was filed, NEPA was passed. Although two years earlier, when GWQCB first issued its order, such a permit would have been granted as a matter of routine, the Army official receiving the request was sensitive to the implications of the new law. As he saw it, the dock construction permit was an essential part of the plan to barge the wastes out to sea. Thus, granting the permit could be construed as a "federal" action that would "significantly affect" the environment. He decided that, by law, an impact statement should be written before the permit could be granted.

The environmental impact statement process focused attention on the problems caused by ocean dumping; it also forced Cyanamid to consider whether there were any alternatives. The disclosures required by the impact statement and the comments by the public and other government agencies revealed that American Cyanamid had not adequately considered the effects of ocean dumping in its initial application. Upon further reflection, American Cyanamid decided that an alternative—building a reprocessing plant that would convert its wastes to a commercially saleable by-product—was more attractive than proceeding with its initial proposal, and it dropped its plans for dumping wastes into the ocean.

Putting NEPA into Perspective

The environmental impact statement process played a slightly different role in each of the foregoing situations. In the case of the bridge across San Francisco Bay, the statement supported vocal opponents of the project, giving them additional evidence to document their position. It also drew sympathetic federal agencies into the controversy. In the case of Cyanamid, the impact statement process drastically altered the course of events. It required public consideration of the implications of ocean dumping and led to the identification and adoption of alternatives. In both cases, relatively minor officials interpreted the law and made important decisions. Had the officials involved not been conscientious in the execution of their duties, or well-informed on NEPA's application to their actions, the results might have been very different.

Government programs do not always work as well as intended, and the environmental impact statement process is no exception. The statute is based on a relatively simple theory of how agencies behave. It assumes that the establishment of a requirement for an impact statement will result in the preparation of a statement; that all statements written will be comprehensive and unbiased; that the development of environmental information on a project will lead to consideration of that information; and that agency officials reach a decision in an atmosphere that is substantially free of political pressure.

Clearly, these assumptions are not always valid. In an attempt to reduce their workload, to expedite the processing of a project or, perhaps, to avoid public scrutiny of it, agency staff may misconstrue NEPA's requirements. They may avoid writing an impact statement when one is required, or they may unreasonably exclude some environmental consequences or alternatives from analysis in the statement. Nor does the impact statement process eliminate the highly political context in which many agency decisions are made.

It does not remove the pressure which government officials are sometimes subjected to, nor has it substantially altered the inertia in many bureaucracies to "carry on as before." Accordingly, there are still instances when compliance with NEPA is more a pro forma exercise than an open investigation of a project's impacts.

It would be a mistake, however, to conclude that NEPA inevitably has little or no effect. The aforementioned examples show that, with active public involvement, NEPA has led to changes in the final decisions on projects. Moreover, many agencies, of their own accord, have cancelled or modified projects because of unforeseen environmental consequences that were discovered during the impact statement process. And today, numerous proposals are dropped as a result of NEPA even *before* they reach the stage of serious consideration, because their sponsors are aware that they are unlikely to withstand the public scrutiny that the statute requires.

In considering NEPA's effect on agency decision making, it may be helpful to think of the statute as operating in two different arenas. Within the agency, at the staff level, NEPA has started to change the manner in which projects are analyzed. Agencies have been forced to hire a substantial number of individuals who are experts on environmental matters and whose sole function is to analyze the environmental consequences of projects. Agencies also have been required to spend large sums of money on research into environmental problems. The milieu within agencies is thus slowly changing. Moreover, because of the impact statement process, federal, state, and local agencies that otherwise would not have authority to officially express their views on other agencies' projects now have that authority and are taking advantage of it.

All of this creates new information on projects and opens additional channels of communication. The new information and additional channels of communication in turn create new pressures on agencies' decision making, which leads to more environmentally sensitive agency decisions. While the inroads into the traditional ways that agencies review projects have, in many situations, been modest, these inroads lay the foundation for even greater changes over the coming years. Many agencies are still coming to terms with their responsibilities under the act.

NEPA also operates in the public arena to bring outside pressures on an agency. Citizen groups sometimes use NEPA as a legal tool. According to the Council on Environmental Quality, 177 projects had been delayed, as of June 30, 1976, because of injunctions issued under the statute. CEQ also reported that, although no projects had been permanently halted because of a NEPA injunction, 42 projects were cancelled after an injunction under the statute had been issued. In each of these cases, subsequent negotiations in light of the injunction led to a decision to withdraw federal involvement in the project. These successes with NEPA as a legal tool have produced a situation where even a credible threat of a lawsuit under the statute is often a deterrent to agency action.

NEPA's greatest effect, however, has been through the public focus it generates on the environmental damage that a project may cause. NEPA has become a powerful political tool. Groups often seek to get an agency's impact statement to present all the major environmental problems associated with a project. They then use these disclosures to generate widespread support for their opposition to a project and to persuade public officials that the project should be modified or halted.

Both the internal and the external pressures NEPA may place on agency decision making should always be kept in mind. In many cases, citizens will have to direct their efforts to both of these levels. If a citizen group can convince an agency's staff during its deliberations on a project that it should implement the spirit as well as the letter of the act, its task in forestalling environmentally damaging decisions is likely to be much easier. If, however, the group is not successful in persuading agency staff of the need for changes in a particular project, it will eventually have to consider using the impact statement as a legal and political tool. In the final analysis, it is individual citizens, armed with knowledge of the impact statement process and with the diligence and concern necessary to ensure that its requirements are met, who are critical to the realization of NEPA's goals.

2

Which Projects Require an Environmental Impact Statement

This chapter deals more specifically with the kinds of government actions that require environmental impact statements and with how an agency decides whether a particular action requires a statement. It also discusses how a citizen can learn about these agency decisions and how he can use federal regulations to determine whether an EIS should be written. Chapter 3, which discusses the steps a citizen should take while a statement is being prepared, should be referred to if the agency has already decided to prepare a statement. Chapter 4, which discusses how to review the draft statement, applies if the draft statement has already been released.

Agencies, Types of Projects, and Numbers of Statements

Relatively few of the countless decisions made each year by the Federal Government will have a sufficiently major impact on the environment to require the preparation of an impact statement. Obviously, issuing a permit to build the Trans-Alaska pipeline or to construct a large dam would require the preparation of a statement. On the other hand, changing the hours during which an office building remains open or adding lights to part of an expressway would not have far-reaching environmental consequences; thus, an impact statement would not be required.

Not every project requires an EIS.

Each year, although innumerable actions are proposed, impact statements are written on only approximately 1,000 of the most environmentally important ones. The majority of the impact statements written concern transportation projects such as highways, airports, and mass transit systems; water resource projects such as dams, stream channelizations, and dredging of waterways; natural resource management projects in parks, wildlife refuges, and forests; and energy development projects such as power plants, transmission lines, mining, and offshore drilling. Figure 1 shows the major categories of projects for which environmental impact statements were prepared from 1970 to 1975.

Just as certain kinds of projects tend to require impact statements, certain agencies most commonly are involved with projects that significantly affect the environment. The Federal Highway Administration, Army Corps of Engineers, Nuclear Regulatory Commission, Forest Service, and several agencies within the Department of the Interior (National Park Service, Bureau of Land Management, Fish and Wildlife Service, and Bureau of Reclamation) write more impact statements than the average agency (see Table 1). Those agencies empowered to build highways and dams, to approve nuclear power plants, and to preserve natural resources must be especially sensitive to the environmental consequences of their actions. Thus, if an action falls into one of the categories shown in Figure 1, or if it is being undertaken by an agency—such as the Corps of Engineers—that writes a large number of statements every year, there is a good chance that drafting an environmental impact statement will be seriously considered.

FIGURE 1   Draft environmental impact statements filed annually, by project.

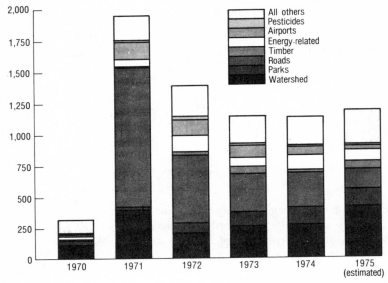

Source: U.S. Council on Environmental Quality. *Environmental Quality—1975*. Sixth Annual Report. Washington, D.C., U.S. Government Printing Office, 1975, p. 640.

Keep in mind, however, that even agencies seemingly far removed from environmental matters must comply with NEPA. NEPA applies to all federal agencies. The Interstate Commerce Commission, for example, prepared an EIS several years ago in connection with its decision on whether to approve an increase in railroad freight rates covering the shipment of waste materials for recycling. It was clear that too high a rate could discourage recycling; this, in turn, could cause environmental damage through increased dumping of paper, cullet, and scrap iron and through increased consumption of natural resources. Similarly, the Food and Drug Administration recently prepared a statement on whether to permit soft drink bottlers to switch from glass and metal to plastic containers. The Department of Defense frequently prepares statements in connection with proposals to close military bases in the United States and transfer base activities to other locations in the country.

While only a few agencies are responsible for most of the impact statements that are written, almost 70 different agencies have written statements on proposed actions at one time or another.

TABLE 1   Draft Environmental Impact Statements Filed by Agency: 1970–1975

U.S. Agency	1970	1971	1972	1973	1974	1975
Agriculture	62	79	124	166	179	189
Commerce	0	8	12	15	12	13
Defense	5	27	24	19	26	14
Army Corps of Engineers	119	316	211	243	303	273
Health, Education, and Welfare	0	1	11	4	0	3
Housing and Urban Development	3	23	26	22	21	78*
Interior	18	65	107	119	109	67
Justice	0	0	3	1	1	0
Labor	0	0	0	3	1	4
State	0	0	3	1	3	1
Transportation	61	1,293	674	432	360	229
Treasury	2	2	5	0	2	0
Energy Research and Development Administration	—	—	—	—	—	7
Environmental Protection Agency	0	16	13	26	14	23
Federal Energy Administration	—	—	—	—	—	5
Federal Power Commission	0	0	38	16	12	29
General Services Administration	3	34	6	24	26	23
Nuclear Regulatory Commission	—	—	—	—	—	26
All others	14	64	63	26	37	13
Total	287	1,928	1,320	1,117	1,106	997

* Total includes 27 impact statements prepared by local governments under the Community Development Block Grant Program.

Source: Adapted from U.S. Council on Environmental Quality. *Environmental Quality—1976.* Seventh Annual Report, Washington, D.C., U.S. Government Printing Office, 1976, p. 132.

The fact that a project is not a dam or power plant does not mean that it will not require an impact statement or that the agency cannot be persuaded that one should be written.

Recent Projects and Agency Decisions on Whether to Write a Statement

Figure 2 uses specific examples to present a clearer idea of when impact statements are needed. It describes two kinds of projects— those for which impact statements have been written, and those for which agencies have decided that statements were not required.

FIGURE 2 Examples of agency decisions on whether an impact statement should be prepared.

AGENCY	PROPOSED PROJECT	ENVIRONMENTAL CONSEQUENCES	WAS IMPACT STATEMENT WRITTEN?
TRANSPORTATION			
Department of Transportation (DOT), Federal Highway Administration (FHWA)	Proposal to upgrade 0.8 mile of an existing road and build a new bridge in Dundy County, Nebraska. Between 2.4 and 10.0 acres would be acquired for a right-of-way. Grading, a gravel surface course, a 1,700-foot long channel change, and culverts would be required.	Water quality of the stream would be temporarily adversely affected; vegetation would be temporarily upset.	Yes Draft EIS filed 3/10/77.
	Proposal to improve Del Monte Avenue in Monterey, California. Phase 1 would widen four blocks of the road. Phase 2 would extend this widening another several blocks. Plans include expanding the roadway and adding medians with turn lanes, new parking, and sidewalks.	Additional traffic would be generated; several existing businesses and residences would be demolished; some temporary construction effects.	Yes Draft EIS filed 11/3/76.
	Proposal to replace a bridge over the Missouri River with two new 4,000-foot bridges and to reconstruct 2 miles of bridge access roads.	Since there is no community surrounding the project area, the proposal would not require relocation of people nor would it disturb community character. During construction, however, there would be a	Yes Final EIS filed 3/9/77.

Project	Impact	Draft EIS filed 2/23/77.
road between three highways in Marietta, Georgia. Project length is 1.5 miles.	tions; permanent increases in noise and air pollution.	
$5.6 million bridge replacement in Oswego County, New York. Two lanes would be increased to four, and the carrying capacity of the bridge would be increased.	Temporary construction-related impacts, such as increased turbidity and noise pollution. No businesses or residences would be relocated. Expected increase in traffic flow and better service to existing businesses.	No
Improvement of 24 miles of New York Route 7 between Troy and the Vermont border.	Present poor highway conditions would be alleviated. Very little displacement of businesses and residences and no disruption of the community are anticipated. Temporary construction-related impacts, such as traffic rerouting, are expected.	No
Proposal to widen 1.5 miles of road from three to five lanes in Niagara County, New York.	Traffic flow, safety, and storm drainage would be improved. Minimal increase in air and noise pollution. Increase in saline pollution, but with minimal anticipated effects on local water.	No
Reconstruction of a 6-lane highway, and rehabilitation of existing park areas listed as eligible for inclusion in the National Register of Historic Places, in Brooklyn, New York.	Temporary rerouting of traffic. Eight inches of park land would be used for the construction of curbs. Substantial decrease in accidents, most of which are presently due to the rough road condition.	No

FIGURE 2 (continued)

AGENCY	PROPOSED PROJECT	ENVIRONMENTAL CONSEQUENCES	WAS IMPACT STATEMENT WRITTEN?
	Proposal to amend FHWA rules to exclude the construction of pedestrian walkways and bikeways from classes of FHWA actions which potentially could require EISs.	The proposed rules also require each applicant for construction funds to certify that all possible planning to minimize harm in the design and location has been accomplished; thus, any environmental damage that might occur would likely be minimal. The rules are not expected to increase the number of applications for walkways and bikeways.	No
DOT, Federal Aviation Administration (FAA)	Proposal to expand the Asheville, North Carolina airport by extending an existing runway by 1,500 feet. Parallel taxiways, land acquisition, construction, and lighting would be necessary.	Temporary construction impacts; increased noise and air pollution.	Yes Draft EIS filed 3/25/77.
	Proposal to amend FAA regulations to provide more stringent noise level requirements for several different types of airplanes.	Decrease in noise pollution. Minor increase in fuel consumption.	Yes Final EIS filed 12/6/76.
DOT, Coast Guard	Proposal to permit the construction of a toll bridge across a navigable waterway (Southern Crossing, San Francisco Bay).	Direct effects would be minimal, but secondary effects would include decreased use of mass transit, increased traffic, and the probable later disruption of shoreline ecology by the construction of	Yes EIS filed 1/5/71.

WATER RESOURCES

Department of Defense (DOD), Army Corps of Engineers

Project	Impact	EIS
Proposal to improve an existing levee system in Okanogan County, Washington; 14,100 feet of levee, including 550 feet of flood wall, would be constructed. Floods with a 1 percent chance of occurrence annually would be prevented.	Total vegetation loss on 19.7 acres; increase in turbidity.	Yes Draft EIS filed 2/23/77.
Proposal to dredge Great South Bay, Long Island (New York) in accordance with an accepted federal navigation project by the removal of 17,600 cubic yards of material from a channel 8 feet deep × 75 feet wide × 1 mile long.	Adverse effect on marine biota.	Yes Final EIS filed 12/24/76.
Proposal to clear and snag debris from a 33-mile stretch of the Pembina River in North Dakota.	Flooding would be reduced on 760 acres of land.	Yes Draft EIS filed 12/1/76.
Proposal to issue a permit for dock construction on a navigable waterway for barge-loading facilities for ocean dumping of toxic wastes.	Minor direct effects from the dock construction, but possibly significant pollution from the unsupervised ocean dumping of manufacturing wastes.	Yes Draft EIS filed 9/6/72.
Proposal to add hydroelectric production to the Lucky Peak Dam in Idaho. Construction of a side tunnel from the existing outlet tunnel would be necessary, as would a new powerhouse.	Temporary construction impacts; possible negative aesthetic impacts from the construction of a surge tank (for storage).	Yes Draft EIS filed 12/28/76.

FIGURE 2 (continued)

AGENCY	PROPOSED PROJECT	ENVIRONMENTAL CONSEQUENCES	WAS IMPACT STATEMENT WRITTEN?
	Proposal to dredge the Ogdensburg Harbor in New York to maintain navigable depth.	Temporary air and noise pollution. Temporary degradation of water quality in the harbor due to increased turbidity and suspended solids. Temporary loss of 10 acres of open-field vegetation. Some visual and traffic impacts expected, but could be minimized by planning. Beneficial impacts include continued navigability of the harbor.	No
	Harbor maintenance proposal to dredge in the Wilson Harbor.	Disposal of 15,000 cubic yards of dredged material in open-lake disposal area. Polluted dredge would be dumped first and covered by unpolluted material. Temporary loss of water flora, and air and noise pollution expected. Since the proposed dredging would increase channel depth to 8 feet, an increase in recreational boating is expected, with the concommitant economic benefits.	No
Department of Agriculture (USDA), Soil Conservation Service	Watershed plan proposal for protection of existing reservoirs, flood prevention, municipal and industrial water supply, and water-based recreation in Dynne Creek, Cleburne County, Alabama.	161 acres of wildlife habitat would be lost; 47 acres of flood-plain forest would be cleared.	Yes Draft EIS filed 2/28/77.

Agency	Proposal	Impact	EIS
Department of the Interior (USDI), Bureau of Reclamation	Proposal to enlarge 3.2 miles of channel in Franklin, Washington. The Esquatzel Coulee Wasteway would be extended 0.9 mile downstream, and 1.5 miles of constructed channel upstream would be abandoned. Five thousand feet of unstable bankway would be sloped and stabilized.	Vegetation reduction; displacement of 85 acres of agricultural land, 10 acres of undisturbed land, and 1,500–2,000 acres of rangeland.	Yes Draft EIS filed 2/23/77.
Environmental Protection Agency	Water quality management plan proposed for El Paso and Teller counties in Colorado. The plan recommends classifications and criteria for streams, projected treatment-facility needs, and implementation programs.	Decrease in groundwater discharged to two streams; increase in groundwater demand; and ecological systems impact.	Yes Draft EIS filed 3/25/77.
	Proposal to revise the criteria for deciding whether to issue ocean-dumping permits. The revisions would require preparation of site surveys and implementation of a monitoring program to detect chronic effects before they became irreversible.	Continuation of ocean dumping affects ocean-bottom topography; chronic build-ups of dumped materials are possible. The revisions would require many present dumpers to find alternative methods of waste disposal, which could adversely affect other parts of the environment.	Yes Final EIS filed 2/8/77.
	Proposal to construct a 1.6 million gallon/day sewage-treatment plant to meet expected growth in the Aspen and Snowmass resort areas in Colorado. The project is not considered a growth stimulant, since it would be accompanied by the closing of older facilities in the area.	No adverse effects are expected, other than some degradation of the quality of downstream water and temporary construction impacts.	Yes Final EIS filed 12/10/76.

FIGURE 2 (continued)

AGENCY	PROPOSED PROJECT	ENVIRONMENTAL CONSEQUENCES	WAS IMPACT STATEMENT WRITTEN?
NATURAL RESOURCES USDA, Forest Service	Proposals for the suppression of the gypsy moth by the use of insecticides in 1977 in conjunction with state agencies.	Possible decrease in related, nontarget species.	Yes Draft EIS filed 1/7/77.
	Proposal to construct a nonprofit fish hatchery in Alaska to help restore the depleted salmon-fishing industry. Access trails, a boat landing, residences, a fish hatchery, and other support buildings would be needed.	Discharge of metabolic and human wastes into the water during construction; loss of the wilderness character of the area.	Yes Final EIS filed 1/24/77.
	Proposal to sell insect-damaged timber by logging an area of 12–36 acres, located in various portions of a 480-acre tract on Naked Island, Alaska; 2.25 miles of access road would be needed.	The roadless and undeveloped nature of the area would be lost, and visual impacts would last for 30–40 years.	Yes Draft EIS filed 3/28/77.
	Proposal to improve forage for Alaskan moose in the Alaskan National Forest by prescribed burnings at 139 sites on a 22,000-acre tract over a 10-year period.	Vegetative changes; loss of some wildlife; temporary air-quality reduction; wildfire hazards.	Yes Draft EIS filed 3/22/77.
	Proposal to amend regulations to permit the use of helicopters in the capture and transportation of wild horses and burros.	Past removal of predators encouraged population increases of wild horses and burros in semidesert areas. Transfer of animals to less-populated areas of their ‾‾‾‾‾‾‾	No

Agency	Proposal	Environmental consequences	EIS required
		from overgrazing. Temporary noise and air pollution expected.	No
	Proposal to cut 107 acres of hardwoods and thin 411 acres in the Allegheny National Forest in Pennsylvania by removing mature, high-quality trees and low-quality seedlings.	The cutting would be visible on one section of publicly traveled road; some erosion due to road improvement expected; a few trees would be damaged. Beneficial impacts include improved deer habitat, better hunting, and improvement of the quality of the residual timber stands.	No
	Design and development of boat launches, picnic sites, and bankfishing and gravelling an access road in the Allegheny National Forest, approximately nine miles west of Marshburg, Pennsylvania.	Some sedimentation expected during construction of the boat launch; air and noise pollution levels are expected to increase slightly; approximately 50 acres of timber would be withdrawn from production.	No
USDI, Bureau of Land Management (BLM)	Transfer to the state of Nevada of an additional 9,000 acres of land under the Fort Mohave Act of 1970, P.L. 86-433. The transfer would occur over a 6-year period.	Environmental consequences of alternative approaches for management of the land.	Yes Final EIS on the initial transfer of land under the Fort Mohave Act filed 1/24/75. Supplement to the EIS on the additional 9,000-acre transfer filed 3/8/77.

FIGURE 2   (continued)

AGENCY	PROPOSED PROJECT	ENVIRONMENTAL CONSEQUENCES	WAS IMPACT STATEMENT WRITTEN?
	Plan to close federally owned desert lands administered by BLM in Southern California to vehicle use.	Decrease in air and noise pollution; decrease in potential for wildlife habitat disturbance.	No
USDI, National Park Service	Plan to allow snowmobile use on county-owned roads at the Pictured Rock National Lakeshore in Michigan.	Minor increase in air and noise pollution; increase in local economy due to higher tourism.	No
ENERGY			
Nuclear Regulatory Commission	Issuance of a permit to operate a nuclear power plant on Long Island.	Clearing of 100 acres of woods; damage to fish by high velocity of cooling intakes; thermal stress on marine biota; and radiation exposure risks.	Yes Draft EIS filed 3/24/77.
	Proposal to permit mining and processing of uranium in Converse County, Wyoming over a 10-year period.	Vegetation and soil disturbances related to construction of processing factory and mining; related noise, dust, and air pollution and topography changes.	Yes Draft EIS filed 12/29/76.
USDI, Bureau of Land Management (BLM)	Proposal for new BLM regulations governing leasing of public lands for coal development and revised Geological Survey regulations governing coal explorations, mining operations, and reclamation on public lands. The changes would	Reduction in the amount of unreclaimed land; increased operator expense; increased consumer prices; closings of small mines.	Yes Final EIS filed 9/16/75.

Agency / Project	Description	Environmental Impact	EIS Required
	require submission of detailed plans for exploration and mining prior to operations, reclamation of mined land as an integral part of operations, and an expansion of federal authority to include private surfaces on federally owned subsurface minerals.		
	Proposal to lease 234 tracts (1 tract = 9 square miles) of outer continental shelf land off the coast of Texas, Louisiana, Mississippi, and Alabama for oil and gas drilling.	Potential oil spills.	Yes Final EIS filed 2/17/77.
USDI, Geological Survey	Proposal to lease 2,100 acres of Crow Indian lands in Montana for 20 years for strip mining.	Land surface destruction; destruction of vegetation and aquifers; degradation of ground and surface water; noise and dust pollution.	Yes EIS filed 12/15/76.
HOUSING Department of Housing and Urban Development	Construction of 761 dwelling units on 93.2 acres in Humacao, Puerto Rico. Plans also call for building a cultural center, parks, playlots, and lakes.	Increase in traffic; removal of vegetation during construction; increase in noise; and increase in use of solid-waste and wastewater facilities.	Yes Draft EIS filed 2/25/77.
	Proposal to rehabilitate a rundown urban area in San Francisco to conserve existing housing and to improve neighborhood quality.	Displacement of 150 households; increase in water and energy use, sewage, and solid waste; possible effects on the socioeconomics of the area because, after renovation, other social classes may move into the area.	Yes Draft EIS filed 1/19/77.

FIGURE 2 (continued)

AGENCY	PROPOSED PROJECT	ENVIRONMENTAL CONSEQUENCES	WAS IMPACT STATEMENT WRITTEN?
	Proposal to construct a senior citizens' center in Pensacola, Florida. The building would have a theater/social hall for 200, game rooms, a solarium, an arts and crafts center, a kitchen, rest rooms, and administrative offices.	Offsite erosion and sedimentation; increase in air and water pollution; and loss of 0.35 acre of ground cover.	Yes Draft EIS filed 9/2/77.
	Proposal to build a 7–8 acre park in Augusta, Georgia, including tennis courts, baseball/football fields, recreation center, swimming pool and bathhouse, tot lots, parking lots, and play areas.	Short-term adverse construction effects from the conversion of an industrial area to park uses.	Yes Final EIS filed 2/15/77.
	Urban renewal proposal to replace abandoned buildings in Newark, New Jersey with a one-story housing development.	Short-term construction-related impacts on soil, water, air, and noise expected; temporary increase in economy; long-term increase in neighborhood quality; reduction in rodent infestation; and removal of an "attractive nuisance" for local children.	No
	Proposal to construct two-story garden-apartment housing for the elderly in East Hampton, New York; 48 units are planned to house a maximum of 85 persons.	Loss of some vegetation on the 8.24-acre lot; temporary construction impacts expected. The high permeability of the subsoil and lack of a community sewage-treatment facility make the possibility of groundwater pollution from septic systems fairly high.	No

DOT, Coast Guard	Proposal to acquire six acres for the construction of 62 units of family housing for Coast Guard personnel in Eureka, California.	Permanent displacement of wildlife; potential increase in erosion and sedimentation; negative aesthetic impacts.	Yes Final EIS filed 12/1/76.
MISCELLANEOUS			
National Aeronautics and Space Administration	Proposal to test, process, and deliver rocket motors.	Temporary air pollution and noise pollution caused by test firings.	Yes Final EIS filed 2/14/77.
DOD, Army Corps of Engineers	Proposal to permit the construction of a major recreational/commercial complex at the port of San Francisco's northern waterfront. The complex would include a restaurant, shops, a marina, a parking garage, and open space/park areas.	Increase in traffic; sedimentation from the construction of breakwaters; air; water, and noise pollution.	Yes Draft EIS filed 12/21/76.
Nuclear Regulatory Commission	Proposal to exclude from licensing procedures personnel dosimeters (safety equipment) containing less than 50 milligrams of thorium (a regulated substance).	Since dosimeter production would be facilitated, the dosimeters would be more readily available, and compliance with radiation standards would be easier; distribution of up to 3.75 grams of thorium into the environment annually; recycling of dosimeter manufacturing substances.	Yes Final EIS filed 2/17/77.
Interstate Commerce Commission	Proposal to permit abandonment of a 3.15-mile segment of the Burlington Northern Railroad between Nifa and West Batavia, Illinois.	The goods and traffic carried by the railroad are so minimal that the necessary transfer to automotive transportation would not significantly increase air or noise pollution on the area's roads.	No

This information may help a citizen decide whether a particular project requires an EIS. He should compare the environmental consequences of the project under consideration with those of the projects listed. If the environmental consequences of the project seem to be greater than for projects that have statements, it is likely that an EIS will be required; similarly, if the consequences seem to be lesser than for projects without statements, an EIS probably will not be needed.

Figure 2 reveals that some of the projects without statements are similar in effect to projects with statements; indeed, some may actually have greater environmental consequences. The two categories are separated not by a sharp line but by a gray area, because it is difficult to make precise numerical estimates of environmental harm. Where uncertainty on the magnitude of the harm exists, government officials use their discretion on whether to write a statement. Experience shows that, when a project falls into this gray area, the agency's decision usually depends more on the extent of public concern than on precise measurement of likely damage. The extent of citizen concern over a project is frequently a decisive factor in determining whether a statement will be prepared for the project.

Thus, the projects in Figure 2 fall effectively into three categories: projects that clearly require statements, projects that clearly do not require statements, and projects that are difficult to categorize.

Two points that are raised in this figure should be emphasized. First, NEPA covers not only specific government *actions*, but also overall government *programs;* there are many cases in which a single statement has been prepared on a whole program. For example, statements have been prepared on the Bureau of Land Management's timber management program, the Forest Service's renewable resources program, the Department of Commerce's coastal zone management program for Washington State, the Soil Conservation Service's emergency watershed protection program, and the Fish and Wildlife Service's national wildlife refuge program. In some of these cases, separate impact statements were prepared on individual actions in addition to the program statement. For example, the Department of the Interior wrote a statement assessing the government's coal-leasing program on a national basis. It also wrote separate statements on the issuance of specific coal leases. Thus, while a project may not in itself affect the environment significantly enough to require the preparation of an impact statement, that project may be part of a larger government program which does have significant environmental effects and on which a statement covering the overall program is required.

Second, it is important to note that sometimes several agencies are involved in a single project. Only one statement is required in such cases, and it is usually prepared by just one of the participating agencies. The agency writing the statement is called the "lead agency," and it is generally chosen on the basis of the degree and timing of its involvement. The lead agency is responsible for writing a comprehensive impact statement which all involved agencies can use in their decision-making processes. Throughout this book, the term "lead agency" will signify the agency writing the impact statement under discussion.

When several federal agencies are involved in a project, the EIS is usually written by one of them—the "lead agency"—and the others provide assistance.

How an Agency Decides Whether a Particular Project Requires a Statement

One of the responsibilities assigned to the Council on Environmental Quality was the development of a set of guidelines for federal agencies to follow when implementing the EIS process. A copy of the current CEQ guidelines (as of summer 1978) appears in Appendix B.[1] The guidelines provide both general criteria that an agency should use in deciding whether an action requires the preparation of a statement and a set of procedures that the agency should follow throughout this decision process. Each agency's own regulations provide more specific and detailed information on how that agency will make its decisions and what procedures it will follow.

The Criteria. The National Environmental Policy Act and CEQ guidelines require that agencies prepare statements when considering "major actions" that "significantly affect the quality of the human environment." CEQ's March 1976 report to the president states that, "[i]n practice, most federal agencies do not decide whether a proposed individual or program action is 'major' under Section 102(2)(c) of NEPA. They focus instead on whether a proposed action will 'significantly' affect the quality of the human environment. If so, an EIS is prepared." The key, then, to deciding whether a statement is required is evaluating whether the project will significantly affect the human environment.

The Procedures. Under CEQ guidelines, an agency considering an action which may significantly affect the quality of the human environment must first prepare an *environmental assessment.* An environmental assessment is usually a brief description of environmental consequences likely to result from the project. The nature of these assessments varies greatly from agency to agency. In some cases, they resemble draft EISs; others are little more than a statement that, in the agency's opinion, an EIS is or is not required, followed by the signatures of agency officials. The contents of an environmental assessment and the criteria which each agency uses in making its judgments during the assessment stage are outlined specifically in each agency's own regulations covering the implementation of NEPA.

If the agency decides to write an impact statement on the basis of

[1] In the summer of 1977, President Carter directed CEQ to revise the guidelines and issue them as formal regulations. At this writing, these new regulations had not been completed. A copy of the proposed regulations, which are currently under consideration, appears in Appendix C. The concerned citizen should write CEQ to request a copy of the final regulations as soon as they are available. The new regulations are unlikely to significantly alter the procedures an agency must follow in determining whether to write an impact statement.

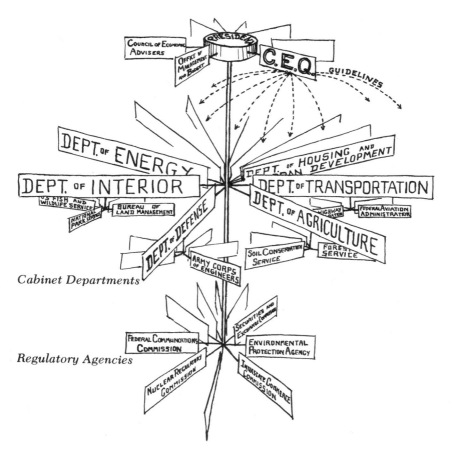

Cabinet Departments

Regulatory Agencies

The Council on Environmental Quality (CEQ) sets guidelines for all federal agencies on the implementation of NEPA.

its environmental assessment, it must then, according to the CEQ guidelines, issue a *notice of intent.* Unfortunately, not all agencies have fully complied with this requirement. CEQ has recently reiterated that all agencies should publish notices of intent in the *Federal Register,* and this procedure will probably be followed more uniformly. A citizen concerned about a specific project, however, can best determine an agency's intention by calling that agency.

CEQ also requires that, if an agency decides not to prepare an impact statement on a project that would "normally" require one,

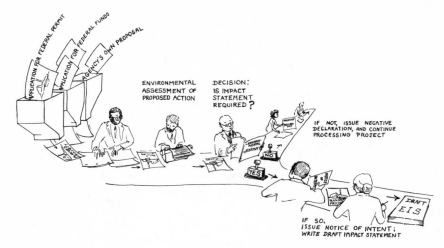

An environmental assessment helps federal officials decide whether they must write an EIS.

the agency must issue a *negative declaration*.[2] This document announces the agency's decision not to write a statement and, in conjunction with the environmental assessment, indicates the reason for such a decision. Again, agencies vary in the degree to which they fulfill this requirement to issue negative declarations.

The Council on Environmental Quality's guidelines require that agencies periodically forward lists of notices of intent and negative declarations to CEQ, and that the notices of intent, negative declarations, and environmental assessments be issued to the public on request. Unfortunately, little of this information is widely circulated. Thus, if a citizen is unable to find a notice of intent, a negative declaration, or other information about a specific agency decision, it is best to *call the agency.*

Other CEQ guidelines require that an agency, after it has issued a notice of intent, prepare a draft environmental impact statement, hold public hearings if there is sufficient interest, allow a reasonable period of time for the public to make comments on the draft statement, and prepare a final statement which incorporates these comments and the agency's response to them. Once again, indi-

---

[2] In CEQ's proposed new regulations, the term "Negative Declaration" has been changed to "Finding of No Significant Impact."

vidual agencies implement these procedures differently; often an agency's own regulations will provide much useful information.

Agency Regulations. Each federal agency, from the enormous Department of Health, Education, and Welfare to the relatively small Canal Zone Government, has developed its own regulations specifying how it will implement NEPA. Agencies have great latitude in their implementation of NEPA, so long as they fulfill the requirements of the statute itself and the CEQ guidelines. Agency regulations establish the particular policies and procedures which *that* agency will follow in carrying out its responsibilities.

Many large agencies issue both general agency-wide regulations and more specific regulations for each major program or bureau. For example, the Department of Transportation has general regulations in addition to separate regulations for its Federal Aviation Administration, Federal Highway Administration, Urban Mass Transportation Administration, and so on.

These regulations outline the criteria and procedures by which each agency decides whether or not it will prepare an impact statement on a given project. Because these regulations vary from agency to agency, a citizen seeking to understand a particular agency's decision should read that agency's regulations. Figure 3 shows the extent to which the NEPA regulations of the major agencies incorporate the basic requirements of the CEQ guidelines.

Almost all agencies have published their regulations in the *Federal Register,* and most of the regulations have been codified in the *Code of Federal Regulations.* Both of these publications are available in large law libraries. Appendix D contains citations to these regulations. Appendix E contains a list of the names and addresses of individuals at federal agencies who can supply citizens with a copy of their agency's NEPA regulations.

FIGURE 3   Adherence of selected federal departments and agencies to the guidelines of the Council on Environmental Quality for environmental impact statements.

ISSUES IMPLICIT IN COUNCIL ON ENVIRONMENTAL QUALITY GUIDELINES	REQUIREMENTS OF COUNCIL ON ENVIRONMENTAL QUALITY GUIDELINES	USDA Department of Agriculture	ASCS Agricultural Stabilization and Conservation Service	FmHA Farmers Home Administration	FS Forest Service	REA Rural Electrification Administration	SCS Soil Conservation Service	EDA Economic Development Administration	COE Corps of Engineers	EPA Environmental Protection Agency	HEW Department of Health, Education and Welfare
Does the agency specify which programs/activities are covered by its procedures?	• Proposed legislation. • New or on-going Federal or federally-assisted projects, programs or permitting activities. • Making of rules, regulations, procedures and/or policy.	Yes	No	No	Yes	Yes	No	No	Yes	Yes	No
Does the agency define a "significant impact?"	NEPA (sec. 101(b)) indicates 7 significant effects to be considered.	No	No	No	No	Yes	No	Yes	No	Yes	No
Does the agency provide for public hearings? Is notice required?	Give criteria on when to hold hearings.	Yes	Yes	Yes	Yes	Yes	Yes	Yes	Yes	Yes	Yes
	Require agencies to use best means for publishing.	Yes	Yes	No	Yes	No	Yes	No	Yes	Yes	No
Does the agency require a negative determination when an EIS is not to be prepared? Is this published?	Lists of negative determinations should be sent to CEQ for quarterly publication in the Federal Register.	Yes	Yes*	Yes	Yes	Yes	Yes*	Yes	Yes*	Yes	Yes
		No	No	Yes	No	Yes	Yes	No	No	No	No

Question	CEQ requirement											
...[require a notice of] intent to prepare an EIS? Is this published?	Agencies should include an early notice system to inform public of the decision to prepare a draft EIS. Revised lists should be sent to CEQ at least quarterly for publication in the *Federal Register*.	Yes†	Yes†	No†	Yes†	Yes†	Yes	Yes†	Yes	No†	Yes	Yes
Does the agency provide for a minimum 45-day review of the draft EIS?	Not less than 45 days.	60	60	60	45	60	60	45	—	45	45	45
Does the agency provide for a minimum 30-day review after the final EIS, before the agency takes action?	Not less than 30 days.	30	30	30	30	30	30	30	—	30	30	30
Does the agency require the 8 sections of analysis suggested by the CEQ?	Requests 8 areas of content analysis.	7	Yes	Yes	7	Yes	Yes	Yes	Yes	Yes	Yes	Yes
Is the agency's format clear, understandable, substantive, and brief?	Ask for clear, understandable, substantive, and brief sections.	Yes	Yes	Yes	Yes	Yes	Yes	Yes	Yes	Yes	Yes	Yes
Does the agency list any actions that are excluded from preparation of an EIS?	Request agencies to identify them, if any.	Yes	Yes	No	Yes	No	Yes	No	Yes	Yes	Yes	Yes
Does the agency list thresholds which mandate preparation of an EIS?	Request agencies to identify them, if any.	No	Yes	No	No	No	Yes	No	No	No	No	No

* ASCS, SCS, COE, and NPS require that lists of the negative determinations be sent to the CEQ quarterly for publication in the *Federal Register*.

† USDA, ASCS, FS, REA, EDA, COE, and NPS provide CEQ with a quarterly list of notices of intent for publication in the *Federal Register*.

‡ FHA, BuRec, and TVA use notices of availability of the EIS, not notices of intent.

FIGURE 3 (continued)

ISSUES IMPLICIT IN COUNCIL ON ENVIRONMENTAL QUALITY GUIDELINES	REQUIREMENTS OF COUNCIL ON ENVIRONMENTAL QUALITY GUIDELINES	HUD (Department of Housing and Urban Development)	BLM (Bureau of Land Management)	BOR (Bureau of Outdoor Recreation)	BuRec (Bureau of Reclamation)	NPS (National Park Service)	TVA (Tennessee Valley Authority)	DOT (Department of Transportation)	FHWA (Federal Highway Administration)	VA (Veterans Administration)
Does the agency specify which programs/activities are covered by its procedures?	· Proposed legislation. · New or on-going Federal or federally-assisted projects, programs or permitting activities. · Making of rules, regulations, procedures and/or policy.	Yes	No	No	No	No	Yes	Yes	Yes	Yes
Does the agency define a "significant impact?"	NEPA (sec. 101(b)) indicates 7 significant effects to be considered.	Yes	No	No	Yes	No	No	Yes	No	No
Does the agency provide for public hearings? Is notice required?	Give criteria on when to hold hearings. Require agencies to use best means for publishing.	No No	Yes Yes	Yes Yes	Yes Yes	Yes No	Yes No	Yes Yes	Yes Yes	Yes No
Does the agency require a negative determination when an EIS is not to be...	Lists of negative determinations should be sent to CEQ for quarterly publication in the Federal Register.	Yes No	Yes No	Yes No	No No	Yes* Yes	Yes No	Yes* Yes	Yes Yes	Yes* No

Question	Standard									
Does the agency employ a standard notice system to inform public of the decision to prepare a draft EIS. Revised lists should be sent to CEQ at least quarterly for publication in the *Federal Register*. (to prepare an EIS? Is this published?)		No / No	Yes / Yes	No / Yes	No / Yes	Yes / Yes	Not / No	Not / No	Yes† / Yes	Yes / Yes
Does the agency provide for a minimum 45-day review of the draft EIS?	Not less than 45 days.	45	45	45	45	45	45	45	45	60
Does the agency provide for a minimum 30-day review after the final EIS, before the agency takes action?	Not less than 30 days.	30	30	30	30	45	30	30	45	30
Does the agency require the 8 sections of analysis suggested by the CEQ?	Requests 8 areas of content analysis.	Yes	Yes	Yes	Yes	Yes	Yes	Yes	9	Yes
Is the agency's format clear, understandable, substantive, and brief?	Ask for clear, understandable, substantive, and brief sections.	Yes	Yes	Yes	Yes	Yes	Yes	Yes	Yes	Yes
Does the agency list any actions that are excluded from preparation of an EIS?	Request agencies to identify them, if any.	Yes	No	No	No	No	No	Yes	No	No
Does the agency list thresholds which mandate preparation of an EIS?	Request agencies to identify them, if any.	Yes	No	Yes	No	No	Yes	No	No	Yes

* ASCS, SCS, COE, and NPS require that lists of the negative determinations be sent to the CEQ quarterly for publication in the *Federal Register*.
† USDA, ASCS, FS, REA, EDA, COE, and NPS provide CEQ with a quarterly list of notices of intent for publication in the *Federal Register*.
‡ FHA, BuRec, and TVA use notices of availability of the EIS, not notices of intent.

Source: Commission on Federal Paperwork. *Environmental Impact Statements.* Washington, D.C., U.S. Government Printing Office, 1977, pp. 24-27.

How a Citizen Can Make
an Independent Decision

Citizens most frequently are interested in the environmental impact statement process because they are concerned about a specific project. If the agency planning to go ahead with the project has decided not to prepare an impact statement, a citizen may want to determine on his own whether the agency has fully complied with NEPA in reaching this decision. A study of the following items will help citizens judge whether or not an impact statement is required.

1. *A detailed description of the project.* Information such as size, location, and likely effects on the populace and the physical environment should be developed. There are no formal mechanisms for obtaining this information; some digging may be required here. Possible sources of information include the involved agency; the applicant, if a federal permit is necessary; state and local environmental agencies; consulting firms; and university professors.
2. *The CEQ Guidelines.* A copy appears in Appendix B.
3. A copy of the agency's NEPA regulations. See Appendixes D and E for information on how to obtain these regulations.

There are three alternative approaches to making an independent decision; each of them should lead to the same conclusion. The approach to use depends largely on which information is most readily available.

If the decision on whether an impact statement is required is close, the citizen may want to use more than one approach to get a clearer picture of the situation. In some cases, it is a good idea to follow through on each of the three approaches as a means of double-checking the conclusion. This procedure also will help the citizen to substantially buttress any argument he might wish to present that an impact statement is required.

The first approach is to compare the project description with the CEQ guidelines and the agency's own regulations. The citizen should begin by checking the agency's regulations to see whether the project falls into a category of projects for which statements are always required or a category of projects for which statements are never required. If it does, the issue is settled. If it does not fall into either category, the citizen should study the regulations to find the specific criteria that the agency official must apply in deciding whether an EIS should be written. By applying these criteria to the description of the project, the citizen should be able to form an independent judgment on the reasonableness of the agency's decision.

Another way to make an independent decision is to look at the types of projects for which the agency has written statements in the past. There are two publications—*The 102 Monitor*[3] and *EIS*[4]— which can be consulted. Issued monthly, they describe every project for which an EIS has been filed with CEQ during the preceding month. Comparing the agency's decision on the proposed project with its decision on similar projects in the past will show whether the agency is fairly and consistently implementing NEPA.

To decide for himself whether an EIS is required, the citizen may need to compare what he knows about the project and its likely effects with the relevant agency's regulations to implement the National Environmental Policy Act. It may be necessary to consult the 102 Monitor.

---

[3] *The 102 Monitor.* Washington, D.C., U.S. Government Printing Office, 1971– .
[4] *EIS: Key to Environmental Impact Statements.* Washington, D.C., Information Resources Press, 1977– .

The citizen may have to scan a dozen or more back issues of *The 102 Monitor* or *EIS* to review enough similar projects to reach a decision. The task, however, is not difficult. It involves looking at a number of project descriptions similar to those summarized in Figure 2, and assessing the project about which the citizen is concerned against them.

The best place to look for *The 102 Monitor* and *EIS* is in large libraries. *The 102 Monitor* can be obtained on a subscription basis, at a modest cost, from the U.S. Government Printing Office. *EIS* is available on a subscription basis from Information Resources Press. It contains more comprehensive descriptions of projects (average 400 words) and is better indexed (subject, agency/organization, geographic area, legal mandates) than *The 102 Monitor*. It is also considerably more expensive.

The last approach to making an independent decision is to apply the basic legal tests. Under Section 102(2)(c) of NEPA, an impact statement is required whenever there is a proposal for major federal action significantly affecting the quality of the human environment. To determine whether this is the case requires an affirmative answer to four subquestions:

1. Is there a "proposal?"
2. Is it for "federal" action?
3. Is the action "major?"
4. Will it "significantly" affect the quality of the "human environment?"

Whether or not there is a "proposal" depends on whether the Federal Government is actively and officially planning to do something. A bright idea in a single individual's mind will not suffice. The project must have advanced far enough so there is concreteness to the proposal and some formal acknowledgement, adoption, or serious consideration of it.

Is the proposal for "federal" action in contrast to state, local, or private action?

The concept of what actions or projects are federal for purposes of NEPA was relatively straightforward until a few years ago. One looked to the overall project, and if the Federal Government provided any essential component of it, the whole project was considered federal for purposes of NEPA. Thus, if a federal permit or license was required before a project could proceed, an impact statement was required even though the project itself was to be wholly funded, constructed, and operated by private groups. This was the case, for example, for the Trans-Alaska pipeline.

Another large group of projects for which impact statements were required resulted from the Federal Government's funding activities. If any federal funds were provided for a project, the whole project became a federal project for purposes of NEPA. This was true of highways partially funded by the Federal Highway Administration (FHWA) or sewage-treatment plants partially funded by EPA.

Finally, there was the group of activities directly and wholly undertaken by a federal agency. These were the clearest examples of federal actions and consisted of such projects as the detonation of a nuclear test blast in Amchitka, Alaska and the Department of Agriculture's decision to spray national forests with a new pesticide. These three groups of activities—licensing activities, funding activities, and activities directly and wholly undertaken by the Federal Government—constituted the traditional groups of federal activities subject to the impact statement process of NEPA. In addition to these major groups, there were a few minor groups, such as proposals of legislation or major policymaking, but they are not sufficiently common to warrant discussion here.

In the last several years, however, the concept of what constitutes federal action for purposes of NEPA has become much more complex. A case decided at the end of 1972 by a federal district court in California highlights the ambiguity. In *Sierra Club v. Volpe*,[5] the court held that reconstruction of a segment of highway was a federal action even though no federal funds were to be used in the reconstruction. The case concerned a portion of State Highway 1 in California. Federal funds were used to expand the northerly portion of the highway, and federal funds were likely to help expand the southerly portion. A portion in the middle—a 6.3-mile segment called the Devil's Slide By-Pass—also was slated for federal funding until environmental groups requested that an impact statement be prepared. According to the court, federal and state officials apparently then decided to avoid the impact statement requirement by foregoing federal funds for this controversial segment.

The court ruled that federal agencies may not circumvent NEPA by segmenting a unified highway project, because it would thwart the purpose of the act. Had the court ruled otherwise, agencies could have divided projects, separating those parts that might result

[5] *Sierra Club v. Volpe*, 351 F.Supp. 1002 (N.D. Cal. 1972).

in significant environmental damage from those that would not. While this decision complicates the concept of what constitutes a federal action for the purposes of NEPA, it safeguards against the possibility that federal officials will try to avoid the EIS process in this way. But because it is no longer as clear which projects are federal, it may be necessary at times to seek legal advice.

Apart from the requirement that there be a proposal for federal action, the action also must be "major" and likely to "significantly" affect the environment before an impact statement must be prepared. The stipulation that a project be "major" means that substantial planning or funding must be involved. While this is slightly vague, the rule of thumb that most agencies and courts follow is that, when a federal action is likely to significantly affect the environment, it is deemed a major action for purposes of NEPA regardless of the amount of money or the magnitude of planning that is involved.

Fortunately, the concept of "significantly affecting the environment" is becoming clearer. According to the CEQ guidelines and court decisions, the "environment" is to be broadly defined and includes the social environment as well as air and water pollution and physical changes in the land.

For example, in *Hanly v. Kleindienst*,[6] the U.S. Court of Appeals for the Second Circuit in 1972 directed the General Services Administration to consider such matters as exposure of neighbors and passersby to drug addicts and the potential for increased crime in the nearby area before constructing a jail in Manhattan. The court said, in effect, that the agency should look beyond such physical effects as generation of garbage or air pollution and, in good faith, attempt to consider what the overall effect on the human environment might be. This generally includes examining potential social as well as physical consequences.

To decide whether a project's effects are significant for purposes of NEPA, the court, in the *Hanly v. Kleindienst* case, suggested two tests. First, one should look to the extent to which the proposed action will cause adverse environmental effects *in excess of* those created by existing uses in the affected area. Thus, the generation of 80 decibels of sound might not be significant in a noisy industrial area, but it could be significant in a hospital zone. One therefore has to look at adverse effects not only in terms of their absolute

[6] *Hanly v. Kleindienst*, 471 F.2d 823 (2d Cir. 1972), *cert. denied* 412 U.S. 908 (1973).

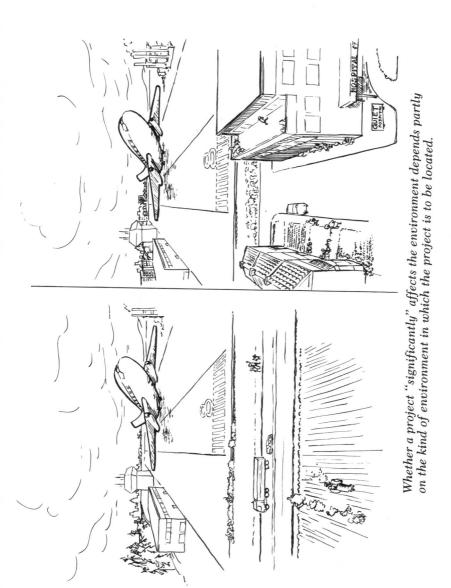

Whether a project "significantly" affects the environment depends partly on the kind of environment in which the project is to be located.

magnitude but also in terms of the existing degree of degradation in the affected area.

The second test was that the agency must look out for "the straw that breaks the back of the environmental camel." Even slight additional pollution in an area already heavily polluted might have a much greater impact than it would elsewhere.

Again, this decision complicates matters somewhat. Applying the legal tests is not always easy to do. In referring to these basic legal tests to decide whether an impact statement is required, someone with legal experience probably should be consulted.

Objecting to an Agency's Decision
Not to Write a Statement

It is an uphill battle to reverse an agency's decision that an impact statement is not required for a particular project. An individual or group attempting this should be prepared to do a lot of work. One or more government employees probably spent a lot of time weighing the competing considerations before reaching the decision; a reversal at this level is thus unlikely. And the decision-maker's supervisor is also going to be hesitant to reverse the agency's decision, for while he is authorized to do so, he probably does not understand the facts as well as the employees under him. He also may be concerned about overturning a decision reached by subordinates and thus demand exceptionally strong reasons before seriously considering any reversal.

Agencies, however, have decided to write an impact statement on a project after initially concluding one was not necessary. Usually, one or both of two developments impelled them to change their thinking. First, new information was uncovered, suggesting that environmental consequences would be much more severe than originally envisioned. For example, a highway may have been planned through the habitat for an endangered species. Perhaps the person deciding to dispense with an impact statement was unaware that the plant or animal life was on the endangered species list. This new information may be sufficient to persuade the agency to reverse its decision.

A second basis for reconsideration has been pressure brought to bear on the agency. This may occur in a number of ways. Other federal agencies may criticize the agency's decision. They may write letters to the staff person—or to his superior. They may even appeal the matter to a high level official in Washington, D.C. Similarly, state and local agencies may criticize the federal agency decision. Public interest groups—or just a few individuals—may criticize the decision and generate substantial controversy over the project. In such situations, the agency may conclude that the wisest course of action is the preparation of an impact statement.

Getting Started. Before starting a campaign to change an agency's decision, a citizen or interest group should get a copy of the agency's negative declaration and the supporting environmental assessment and find out *why* the agency concluded that no statement was required. In order to get the agency to reverse its decision, it must be demonstrated that the agency didn't have all the facts when it decided a statement was not required or that the conclusion that it drew from the facts was clearly in error.

Uncovering New Facts About the Project. Before starting to dig for new facts, the citizen or group should decide what environmental consequences are of greatest concern. Is it the additional traffic? The water pollution? The loss of open space? Pinpointing the possible problems will help the citizen focus his efforts.

The citizen should then decide what kind of information might help shed some additional light on the project. Visits to other federal, state, and local agencies may be useful. These agencies not only may be able to offer suggestions, but may have some of the exact information the group needs.

A visit to the project site may uncover additional information, and consultation with faculty at a local university also may be helpful.

There is no general watchword other than *DIG*. The citizen is searching for facts that the agency was unaware of, which suggest that the environmental consequences of the project may be much greater than the agency thought.

Putting Pressure on the Agency. There may be no additional facts. The agency, however, may have unreasonably concluded from its information that the project would not significantly affect the environment. In such a situation, pressure should be put on the agency. If there are additional facts, the task will be somewhat less difficult, although by no means is it likely to be easy.

Citizens should first arrange an appointment with the appropriate agency official to discuss the decision. It is important to give the agency a chance to reverse a decision on its own before mounting a campaign. At this meeting the citizen or group should outline the reasons in favor of preparing an impact statement. References to the agency's regulations, the CEQ guidelines, and the agency's past decisions on similar projects should be included in the discussions.

If the agency is not convinced that it should change its decision, the citizen or group must then organize a campaign to convince others (different agencies, interest groups, and politicians) that the agency made a mistake and that the agency's position must be altered. While campaign organizing is beyond the scope of this book, here are a few suggestions:

1. A letter should be written—not more than a few pages long—stating why the project is likely to produce substantial environmental damage. The objective of the letter is not to villify the erring agency but to convince the person reading it. The letter should be sent to CEQ, to the nearest EPA office, to the state environmental agency, and to the newspapers and should be addressed to the appropriate person by name. At the end of the letter, a request for assistance should be made. When letters are written to newspapers, it is advisable to ask that the letter be published and that the newspaper write an editorial.

2. Each of the letters should be followed up—by telephone, if necessary, but preferably in person. In this way, it is possible to explain again to the person to whom the letter was sent what the group's concerns are and to ask for his support.

3. It is often helpful to contact local environmental groups and public interest organizations, such as the League of Women Voters, to find out if they are aware of the project. If not, they should be given a description of the project. Presentations to local groups urging them to ask the agency for an impact statement may be appropriate.

4. Local officials, state representatives, and congressmen and senators should be contacted. The situation should be explained to them, and they should be urged to ask the agency for an impact statement. It is important to let politicians know that not just an isolated individual but a large number of their constituents are concerned about the project.

5. As a last resort, the group may have to consider filing a lawsuit. Although litigation is expensive, it may be the only way to get the agency to alter its decision. While a local general attorney may be able to handle the case, a lawyer who specializes in environmental problems may be preferable. He is probably already an expert in

the applicable law and may be able to more easily mount an offense. Public interest law firms such as the Environmental Defense Fund, the Sierra Club Legal Defense Fund, and the Natural Resources Defense Council occasionally will take on a case with only nominal financial support from individuals in the community.

Timing of the Preparation of the Statement

The timing of the preparation of the impact statement is crucial, because it substantially affects the extent to which the results of the NEPA process are incorporated into the planning of the project. If the project planning is essentially completed before the impact statement is written, experience shows that modifying the project will be difficult, regardless of the problems that are subsequently uncovered. At this late date, all of the economic, technical, legal, political, and other pertinent considerations will have been meshed into an overall proposal. The sponsors are likely to resist changing the proposal because of a desire to preserve the balance that has been struck among competing considerations, frequently only after prolonged study and negotiations, and because of a desire to avoid delay and additional costs in the implementation of the project. Thus, it is almost essential that the preparation of an impact statement occur early in the planning of a project.

Although there is no rule that covers all situations or that specifies exactly when in the planning of a project the impact statement should be written, the idea is clear: the environmental analysis should be performed concurrently with the economic, technical, and other basic analyses. In most cases, it will be desirable for statements to be written at the feasibility analysis (go–no go) stage rather than at the engineering design stage. Several agencies' NEPA regulations establish a point by which the final impact statement should be completed; however, in many of these cases, the point is beyond the time when the results of the EIS process could be most effectively incorporated into the consideration of basic alternatives. In 1977, the U.S. General Accounting Office (GAO) issued a report in which it concluded, on the basis of a study of 29 projects of the Environmental Protection Agency, Army Corps of Engineers, Federal Highway Administration, and General

Citizens may need to urge government officials to begin the EIS process earlier than the agencies routinely do. For most projects, the EIS should be prepared at the beginning of the planning process, when changes can be made more easily.

Services Administration, that impact statements were generally prepared too late in the development of projects. According to GAO:

[If impact] statements are prepared while projects are being planned, the environmental, economic, and technical factors can be considered together. The impact statements can be completed in time to accompany the project proposals through the agency review process, as intended by the act.

If impact statements are not prepared at the proper time, agency officials may have to choose between delaying a project while completing the statement or advancing a project before statement completion to avoid a project delay. If postponed too long, statement preparation can become a perfunctory task and the document of little use in agency decisionmaking.

Procedures or practices in the Corps of Engineers, Environmental Protection Agency, and General Services Administration permitted late statement preparation. Two kinds of situations resulted. First, two Environmental Protection Agency projects—both treatment plants—were delayed about 10 to 17 months to prepare a statement before starting construction. Second, environmental impact statement preparation in 11 cases lagged behind one or more of the three stages of project decisionmaking—planning, design, or construction. While these 11 projects proceeded without delay, decisions were made without the benefits of an environmental impact statement.

RECOMMENDATIONS TO AGENCIES

The Secretary of the Army should monitor the Corps of Engineers' practices, and the Administrators of the General Services Administration and the Environmental Protection Agency should revise their procedures so that environmental impact statements are prepared concurrently with project planning and completed in time to accompany proposals through agency review processes for approval.[7]

CEQ, in its proposed new regulations to implement NEPA, directs each agency to integrate the impact statement process with its other early project planning. According to the council, this is necessary to ensure that planning and decisions reflect environmental values, to avoid delays later in the NEPA process if unforeseen environmental problems arise, and to head off potential conflicts. When finally adopted, these regulations will have the force of law, and will require much earlier preparation of statements.

The concerned citizen, in his discussions with an agency on a particular project for which an impact statement is necessary, should ask its staff when the agency plans to prepare and release the draft statement. If this is later than the completion of corresponding other studies, the citizen should object and urge that the date for beginning the NEPA process be advanced. Certainly, the spirit of injecting environmental values into the planning and decision making on projects demands that an understanding of a project's environmental impacts be available during the stages when that planning and decision making are being carried out.

A special problem arises with projects planned by private parties, such as electric power plants, or by state or local governmental agencies, such as sewage-treatment plants. For these projects, it is difficult to begin the NEPA process until a complete application,

[7] U.S. General Accounting Office. *The Environmental Impact Statement—It Seldom Causes Long Delays But Could Be More Useful If Prepared Earlier.* Washington, D.C., 1977, pp. i–iii.

including the preliminary project design, has been filed with the appropriate federal agency. In such cases, according to the proposed new CEQ regulations, federal agencies shall require the sponsors of the proposals to initiate the environmental studies. The regulations direct federal agencies that become involved late in the planning of a project to consult with interested private persons and organizations at the beginning of the agency's review of that project and to commence the NEPA process immediately after the application is received.

3

Steps Citizens Can Take Before the Release of the Draft Impact Statement

This chapter discusses the steps a citizen should take to prepare for the release of a draft statement after an agency has issued a notice of intent. If the draft statement has already been released and this information is not needed, skip to Chapter 4, which discusses how to review the statement.

It may take an agency months or even years to release a draft impact statement after it has decided to prepare one. The Departments of Health, Education, and Welfare, Housing and Urban Development, and Commerce generally take three to four months to write a statement. The Federal Highway Administration takes an average of 10 months, the Bureau of Land Management, 20 months, and the Soil Conservation Service, 48 months (see Table 2). The concerned citizen should not view this time as a "waiting period." Much important work can and should be done at this stage.

Joining or Creating a Group

Reviewing an environmental impact statement can be a long, tedious, and difficult endeavor if undertaken without help. A person working alone will have to read carefully the entire document or long sections of it, prepare comments, and then follow through to ensure that the agency has given adequate consideration to the suggestions offered. It is much easier to review a statement and to influence the agency's preparation process as a member of a group.

57

TABLE 2   Time Required for Draft EIS Preparation, Fiscal Year 1975 (in months)

	Minimum	Maximum	Average
Agriculture			
Forest Service	1	24	13
Soil Conservation Service*	36	60	48
Commerce	1	5	3
Defense	2	24	5
Air Force	3	12	4
Army	5	15	6
Navy	2	24	4
Corps of Engineers	2	24	9
Health, Education, and Welfare	1	6	4
Housing and Urban Development	3	6	3
Interior	1	18	10
Bureau of Indian Affairs	4	6	5
Bureau of Land Management	2	38	20
Bureau of Outdoor Recreation	4	12	7
Bureau of Reclamation	8	28	19
Fish and Wildlife Service	3	12	8
National Park Service	12	24	14
Geological Survey	12	24	15
Justice			
Law Enforcement Assistance Administration	3	9	7
Labor	5	12	5–6
State	NA		
Transportation			
Federal Aviation Administration	1	9	7
Federal Highway Administration	5	16	10
Treasury			
Energy Research and Development Administration	9	13	11
Environmental Protection Agency	1	13	9
Federal Energy Administration	2	5	3
Federal Power Commission	9	30	15
General Services Administration	4	11	5
Nuclear Regulatory Commission	3	26	10

NA = Not available.
*Includes project planning, not just EIS preparation.

Source: Council on Environmental Quality. *Environmental Impact Statements: An Analysis of Six Years' Experience by Seventy Federal Agencies.* Washington, D.C., U.S. Government Printing Office, 1976, p. 29.

Unfortunately, EISs tend to be overly technical. Interest groups, however, often include experts in different fields, and, even if they do not, groups are usually more successful than individuals in finding experts. Group members also enjoy the benefits of being able to discuss questions and issues with each other, and, perhaps most importantly, they are able to focus their energies on one part of the review process and thus become far more effective than if they had to do everything themselves. Thus, it is almost essential that a citizen join or form a group when seeking to successfully influence an agency's decision.

Joining an Existing Group. If possible, a citizen should join a group that is already interested in reviewing the impact statement instead of trying to form a new group. An established group may have members who are acquainted with the impact statement process, and they may already have contacts with government agencies and experts willing to act in an advisory capacity. Further, by joining an established group, the citizen will likely be able to spend less time organizing a group and more time working on the draft statement.

The choice of a group depends, of course, upon the individual's concerns. For the citizen interested in a project's impact on the environment generally, participating in a group with a broad range of interests and activities and a local chapter, such as the Sierra Club, might be the best choice. For those concerned with a project's impact on a specific aspect of environmental quality, such

Reviewing an EIS can be an awesome task for one person.

as water pollution, joining a group sharing this concern would be the best alternative. Public interest groups represent a great number of special concerns, ranging from the American Pedestrian Association to the Wild Turkey Society; the best thing to do is narrow the choices and find the group with the most relevant interests.

There are several sources of information that may be useful in locating a public interest group that would be willing to help review an impact statement. A local environmental management council may be able to help. The county representatives for the state cooperative extension also may have ideas. Public interest group directories, such as the *Conservation Directory*,[1] are likely to be available in any local library. The Southwest Research and Information Center publishes *The Workbook*[2]—an excellent monthly catalog listing sources of information on environmental, social, and consumer problems. Each issue usually includes an update of addresses of organizations that may be able to help concerned citizens.

Finally, if an EIS has already been released and a copy is available, the section that contains a listing of who received the statement for review should provide the names of organizations that would be willing to help review the EIS. But at this point, after the draft has been released, time is of the essence. If the review is not already underway, it must be started immediately.

If, when making contacts, a group finds other groups with a strong interest in the EIS, a coalition could be formed. The influence of a successful coalition can be impressive because of the political strength in numbers. But a word of warning: A group can weaken its position by aligning itself with a poorly organized group, or one that is considered "irresponsible," and thus become much less effective.

Forming a New Group. As a last resort, a citizen can form his own group to review an EIS. This should be considered only when no established group is willing to undertake the task, for forming a group is, in itself, a time-consuming operation.

---

[1] National Wildlife Federation. *Conservation Directory.* Washington, D.C., 1978. (Revised annually.)
[2] An annual subscription to *The Workbook* can be obtained by writing to the Southwest Research and Information Center, P.O. Box 4524, Albuquerque, New Mexico 87106. Single copies are available for $1.25.

Making Contacts

After organizing for the review of the draft statement, the group should make contacts with appropriate officials in government agencies and with private individuals who will provide technical assistance. These contacts are vitally important, because they will improve the quality of the review and increase the likelihood that any comments will be seriously considered and acted upon by the lead agency.

Agency Contacts. Contacts should be made with appropriate officials in the lead agency before the EIS is released. Thus, upon learning that an EIS is to be written, the citizen should contact the agency to determine who will be responsible for preparing the statement. Appendix E provides the names and addresses of people in each federal agency who should be able to provide this information.

It is important to establish a good relationship with the agency at this point. Unfortunately, polarization often develops once an EIS is released. A little politeness and consideration at this time will help reduce the chances for hostility or tension later.

If possible, citizens should personally visit the individual responsible for writing the statement to ask about the project's description and purpose. This meeting can provide a basic understanding of the agency's conception of the project and an idea of how it plans to approach the EIS. It is also a good idea for citizens to ask for a copy of the agency's EIS regulations and for a copy of one of their old EISs. Experience shows that agency personnel can be extremely helpful, especially when they do not feel that they are being challenged; so arguments over issues should be avoided at this time. The purpose of the initial meeting is to gain a sense of the lead agency's views and to develop a good working relationship with those who will write the statement.

Contacts with other agencies having expertise in areas relevant to the project and to the EIS also are useful. Generally, these agencies assume a more objective perspective on the proposed project and EIS than the lead agency, and they can be quite helpful by offering advice and explanations. At times, citizen groups can aid these agencies by informing them of issues or controversies involving the project, about which they might be unaware. The more agencies and groups who can be persuaded to review the issues included in the draft EIS, the more responses the lead agency will receive and

Contact should be made with the "lead agency" person who is actually writing the draft EIS.

the more thorough its consideration will be of the environmental impacts of the project.

There are several ways to determine which agencies might have an interest in the action. If the draft has been released, the list of interested parties it contains can be consulted. Contacts are most helpful, however, before the draft has been finished. A directory of state agencies will generally provide a good indication of the state agencies' areas of expertise. Appendix II of the CEQ guidelines (Appendix B) lists all federal agencies and their corresponding areas of environmental expertise.

Technical Help. If a group is concerned about the more general aspects of a project, such as how it will look or whether it will disrupt the community, then the group members themselves will likely possess all the expertise necessary to argue their points effectively. After all, as local citizens, they know the community and what type of development they feel is appropriate.

If a group is planning a more thorough review of an EIS, however, which includes an examination of issues such as the project's likely effects on air and water quality, housing patterns, or the local flora and fauna, technical assistance may well be essential.

The questions that are directed at an expert should be as specific as possible, so his time is not wasted and so his answers are directly relevant to the problem. An expert, given essential background data, should be able to provide valuable information concerning the nature and causes of an impact, its consequences, and possible ways of reducing its severity. Each expert, however, will only be able to provide assistance in his particular area. It is therefore important that, before any expert is contacted, the group spend some time discussing the critical issues associated with the project to determine what kind of technical information is needed.

As soon as a group has determined the particular areas in which it needs assistance, it should seek out individuals with expertise in those areas and try to interest them in the project. While these areas may be defined before the EIS is released, specific questions might have to wait until the EIS is available.

In seeking individuals with expertise, a group should contact universities and their affiliated research centers, the local planning board or agency, and other public interest groups which are potential sources of specialized advice. The cooperative extension service and the local environmental management council, among others, can provide more general assistance.

Obtaining the assistance of technical experts can be one of the most important parts of the review process—unfortunately, it also can be one of the most difficult. What follows is a hypothetical account, based on several actual experiences, of one group's efforts to enlist technical help in the review of a draft EIS.

One can go to an expert with specific questions or to get help in uncovering a general problem. But whatever the approach, its success requires knowing what one wants the expert to do.

In order to eliminate an unsafe and confusing traffic intersection at the junction of two highways in a small upstate town, the New York Department of Transportation proposed a plan that called for widening one of the roads and enlarging an access bridge. Because funds from the Federal Highway Administration would be involved, a draft EIS was written and circulated in the community.

While the proposal, if carried out, would certainly have eliminated many of the traffic problems, some community members felt that the EIS had not considered all of the possible side effects of the construction. They also were concerned about the change in the character of the area as a result of the destruction of several businesses and the substitution of an expanded highway.

A local citizen group, the Committee to Preserve the Environment (CPE), decided to review the draft statement. Their first step was to search for technical assistance.

One group member looked through the *Conservation Directory* in the hopes of finding another group that might be willing to help. All of the groups contacted were sympathetic, but were overburdened with projects of their own and more interested in projects of national significance than those of local or regional importance. One person suggested, however, that CPE consult the *Environmental Group Directory*[3] for the upstate region. These groups turned out to be much more helpful, because they were concerned about the region and had experience in dealing with government officials in the area. They gave CPE several specific recommendations on the best way to approach the review.

Another CPE member tried to interest other government agencies in commenting on the draft EIS. He was uniformly unsuccessful until he found out, while reading his college's alumni news, that one of his classmates was a regional administrator for the Environmental Protection Agency. With this friend's assistance, CPE succeeded in interesting the Environmental Protection Agency in engaging in a detailed review of the project proposal and making a formal presentation of its findings to the Federal Highway Administration (FHWA).

A third CPE member tried to enlist the aid of faculty members at the local university. Because her husband was a professor, she knew many faculty members personally, and she was able to convince several of them to volunteer their services. She also contacted

---

[3] Copies of the *Directory* can be obtained from the U.S. Environmental Protection Agency, 26 Federal Plaza, New York, N.Y. 10007.

professors she did not know, but they were far less responsive. Some hinted that they wanted to be paid; others said they might help, but never responded when they were sent information.

Although it wasn't easy to obtain, the technical help proved to be quite valuable. An engineering professor found statistical errors in the traffic counts that were used in the analysis of traffic patterns. A limnologist found that the runoff from the construction would have a serious effect on the ecology of a nearby lake. And an urban planner found that FHWA had overlooked the potential effects of the new bridge on a residential neighborhood. While the question of whether or not to proceed with the project still has not been resolved, the help of experts was invaluable to CPE in getting the FHWA to take a closer look at the highway proposal.

Unfortunately, groups are not always as successful in lining up technical assistance. CPE found much of its help by chance (as in the case of the college friend who worked for EPA) or because its members were acquainted with technical experts (the professor's wife). Thus, one can only advise groups seeking help to explore all possibilities and not get discouraged too easily.

Influencing an Agency's Approach to Preparation of the EIS

Substantial effort is required to influence an agency's EIS preparation process, but the results can be dramatic. Because final decisions on the contents of the draft have not yet been made, an agency may be persuaded to address additional issues. And because all parties concerned are not as polarized and defensive at this stage as they may become after the EIS has been released, communication can be easier.

The group should first try to obtain information from the agency on what it plans to cover in the draft EIS. Then (and this is where the effort is required) the group must perform its own analysis of the proposed action and try to find any additional issues that require study. This can be an extremely difficult and time-consuming process, because the group members and the technical experts who will be aiding them are unlikely to have a detailed description of the proposed action to use in their analysis. Such a description often will not be available until the EIS has been released. Further, the experts will not have the advantage of being able to study the agency's analysis. They will, in effect, have to prepare their own outline of an EIS, without even a completed plan of the proposed project. Still, a group with sufficient resources—of time, money, and expert help—should consider giving this approach a try, because the advantages can be great.

After the group has reached a preliminary conclusion on what it thinks should be included in the EIS, it should arrange a meeting with those who will be writing the statement. Agency contacts the group has previously made will be valuable here. The proposed outline of the EIS which the group has prepared should be carefully compared with the agency's outline to make certain that all the important issues will be covered. If the agency does not plan to address some of the issues, the group should find out why. If the agency's reasons are not persuasive, the group should explain why it thinks that the issue should, nonetheless, be addressed. While the group may not be wholly successful, an earnest attempt should be made to reach mutual agreement on what should be covered in the impact statement.

Public Meetings. A public meeting is another way for a group to communicate its views on a proposed project to an agency. Agencies often schedule meetings during a project's planning process to solicit information from the public. Even though these meetings usually concern the project in general, and not just the EIS, they provide an opportunity for citizens to voice concerns about the upcoming environmental analysis. Information about these meetings usually is published in local newspapers, and sometimes an agency also attempts to contact and inform active public interest groups in the area. If a public meeting has not been scheduled, an agency may be persuaded to hold one, if community interest is high.

The group should attend these meetings, even if it has already been in communication with a member of the EIS preparation team. That member will have had a chance to relay the group's concerns to others in the agency. By attending, however, the group can learn how others view the project and voice the group's concerns publicly. Active participation in public meetings is one of the best ways for a group to make the community aware of important issues, and to marshal support for the group's positions.

Advisory Committees. When considering large-scale or very controversial projects, some federal agencies, such as EPA, have formed advisory committees composed of representatives from other agencies and from public interest groups. These committees meet regularly with the lead agency and its consultants. Because these committees can have a great effect on the impact statement process and, in particular, on the scope and content of the impact statement, working with them is an excellent way for citizens to influence the EIS preparation process. To become involved, a group need only ask the agency if it will be using an advisory committee and, if so, request that a representative from the group be considered for appointment.[4]

Learning the Procedures

If a citizen is going to deal directly with agency personnel in charge of the EIS, participate on an advisory committee, or simply attend public meetings, it is crucial for him to know when and how to deliver information to the agency so it will have maximum impact. Most agencies' EIS regulations are listed in the *Federal Register* and the *Code of Federal Regulations*; these regulations outline each agency's approach to the EIS process and allow an individual to determine the most effective ways in which to communicate his views to the agency.

These regulations also specify how agency decisions should be appealed. For example, DOT regulations stipulate that a request to reverse its decision not to hold a public hearing should address the following issues: the size of the project, the possibility of significant environmental impact, the amount of public interest or controversy over the project, and the quantity and quality of new information likely to be received by DOT at the hearing. A cogent argument incorporating these points would exert great pressure on DOT to hold the public hearing. Being thoroughly familiar with an agency's regulations can be of tremendous help in successfully accomplishing specific goals.

Sometimes, knowledge of an agency's informal operating proce-

---

[4] CEQ's proposed new regulations require each agency to hold a "scoping meeting" in connection with the development of an impact statement. The meeting is to be scheduled by the lead agency, and invitations are to be extended to both other government agencies and members of the public. The purpose of the scoping meeting is to endeavor to reach agreement among all interested parties on the issues that should be addressed in the impact statement.

dures can prove even more useful than an understanding of its formal EIS regulations. For example, it could be quite helpful to know that, while the Army Corps of Engineers delegates full responsibility for the preparation of an impact statement to its regional offices, the Soil Conservation Service edits and approves all of its final statements in its field office in Broomall, Pennsylvania. Agencies also may routinely or occasionally rely on outside consulting firms for impact statement preparation. Unfortunately, there is no reference that outlines each agency's operating procedures; this information can be obtained only through interaction with an agency. An understanding of both the formal and informal procedures can greatly enhance an individual's effectiveness in dealing with an agency; therefore, citizens are urged to pay close attention to these details.

Obtaining the EIS

Because rapid response is critical in the review of a draft impact statement, groups awaiting a particular EIS should ask to be added to the agency's EIS mailing list prior to the release of the draft statement. The EIS should arrive within a week from its release date. If it does not, or if the draft has already been released, the group should call the agency and request a copy. Writing should be avoided, as it will lead to unnecessary delays. Interested citizens should continue to phone until they have reached someone who will send the draft EIS. This is the only way to be assured of a prompt response.

4

Reviewing
the Draft
Impact Statement

Impact statements can be difficult to review. They are often long and may be poorly organized. They sometimes fail to discuss complex issues thoroughly. And they sometimes assume too much technical knowledge on the reader's part. This chapter discusses how to review a draft statement and offers suggestions on how to judge whether it is complete and accurate. It also discusses how to communicate most effectively the results of the review to the agency.

Purpose of the Review

Impact statements are, for the staff of government agencies, a decision-making tool. They alert the staff to the environmental consequences of a proposed project *before* a decision is made on whether to go ahead with the project. Under NEPA, government agencies are required to "look before they leap."

In this sense, the purpose of reviewing a draft statement is to advise the agency on whether its environmental analysis is accurate and complete. Unfortunately, impact statements are frequently defective. For example, a statement for a proposed highway may be inaccurate because the authors of the statement did not fully understand the role in the community of businesses in the path of the highway. The statement may be incomplete because its writers did not discover in their survey of the area that the proposed

69

corridor for the highway goes through an area that periodically floods. One major purpose, then, for reviewing a draft statement is to help ensure that the environmental analysis used in the government's decision making fully and accurately discloses all the environmental consequences associated with the proposed project.

The impact statement process also serves another function, other than providing agency staff with an understanding of the environmental consequences of a proposed project: It provides an opportunity for concerned citizens to influence the government's decision on the project. Through its solicitation of comments, it provides an official channel for citizens to express their views on the acceptability of the environmental consequences. It lets them, in effect, cast a vote on whether the advantages of the proposed project outweigh its disadvantages. While the final decision on the project is the result of an alchemy that is more complicated than a simple counting of the "pro" and "con" comments, all of the comments must accompany the proposal for the project through the agency's decision-making process. A large number of comments expressing opposition to the project often will produce modifications to the project—and sometimes will halt it entirely.

Thus, there are two basic reasons for reviewing a draft impact statement. The first is to ensure that the environmental analysis discloses all the relevant environmental considerations associated with the project. The second is to let the agency officials who will make a decision on the project know whether the public thinks the environmental consequences are acceptable—whether, in the view of those who will be most affected, the benefits of the project outweigh its costs.

Organization of the Impact Statement

All impact statements must contain a description of the proposed project; an analysis of the beneficial and adverse environmental consequences; and an analysis of alternatives to the project, including the alternative of no-action, and a discussion of the environmental consequences associated with each of the alternatives. These sections form the basic elements of the statement.

In addition, because of the specific wording of NEPA, impact statements must cover two other topics which, in most cases, are of only secondary importance. These are the relationship between local, short-term uses of the affected area and the maintenance and enhancement of long-term productivity; and any irreversible and irretrievable commitments of resources which would be involved if the project were carried out.

While all statements must cover these five topics, the organization of impact statements varies from statement to statement. They all contain a summary sheet at the beginning, and almost all have a table of contents. From there on, however, they differ, depending primarily on the specific agency that wrote the statement and the particular project covered.

A sample summary sheet and table of contents for a statement are shown in Figures 4 and 5. This statement was prepared by the U.S. Coast Guard in connection with applications for bridge permits. The permits were needed by the state of Louisiana for part of a 10-mile limited-access expressway around New Orleans which the state proposed to build. Note that the summary sheet contains the name, address, and phone number of the person to be contacted for further information about the statement at the Coast Guard office in New Orleans. Most summary sheets contain information on the agency staff member to contact for further assistance. Note also from the table of contents that the statement covers more than just the bridges, even though only bridge permits were being requested. The statement covers the entire 10-mile expressway. In the spirit of NEPA, the Coast Guard looked at the overall project being planned and asked what all of the environmental consequences would be if permits for construction of the bridges were issued.

The Initial Review

The examination of a draft EIS should begin with a review of the project description, usually located at the beginning of the statement. Here, the reader will be able to determine where the project is to be built and its major features. The reader should then turn to the section of the statement on "adverse environmental effects which cannot be avoided." This section will contain a summary of the harmful environmental consequences of the proposed action.

Taking these two simple steps will enable a reader to quickly gain an overview of the proposed project. It will enable him to briefly review a book-length statement in 10–20 minutes. Even if the reader is sure that he will want to review the entire statement in detail, this procedure should be followed. It lets the reader subsequently focus on specific environmental problems without losing perspective on how they relate to the project as a whole.

FIGURE 4   Sample summary sheet for an environmental impact statement.

SUMMARY

WEST BANK EXPRESSWAY
JEFFERSON PARISH, LOUISIANA

1. *Type of Document:* Draft Environmental Impact Statement

2. *Type of Action:* Administrative

3. *Lead Agency:* Department of Transportation, U.S. Coast Guard

4. *Responsible Office:* Mr. Joseph Irico
   Chief, Bridge Section
   Eighth Coast Guard District
   Hale Boggs Federal Building, Room 1140
   500 Camp Street
   New Orleans, Louisiana 70130
   Phone: AC 504/589-2965

5. *Description of Proposed Action:*

   The proposed action is a project of the Louisiana Department of Highways to upgrade the existing Westbank Expressway in Jefferson Parish, Louisiana, to a limited access expressway. Purpose of the project is to alleviate the existing and projected traffic congestion. As shown on the accompanying map, the expressway route extends from U.S. Highway 90 on the west to Terry Parkway/General DeGaulle on the east. It follows the existing corridor which is 9.7 miles long with a 300 foot wide right-of-way. No new right-of-way is required except for 14 acres of vacant land at the western end for the U.S. 90 interchange.

   Construction will involve the expansion of the existing expressway throughout from four lanes to six (three in each direction), and elevation of that portion from Barataria Boulevard across the Harvey Canal to Terry Parkway/General DeGaulle with erection of twin high level bridges over the canal. Interchanges, with grade separations, will be proposed at all major cross streets. Parallel service roads will be widened from two lanes to three lanes each and converted from two-way to one-way operations. The Harvey Canal Tunnel will be retained for local traffic use with the service roads.

   Federal action involved is the issuance or denial of a Coast Guard perm.. for the high level bridge crossings and associated approaches. No other federal action is involved for this state funded project.

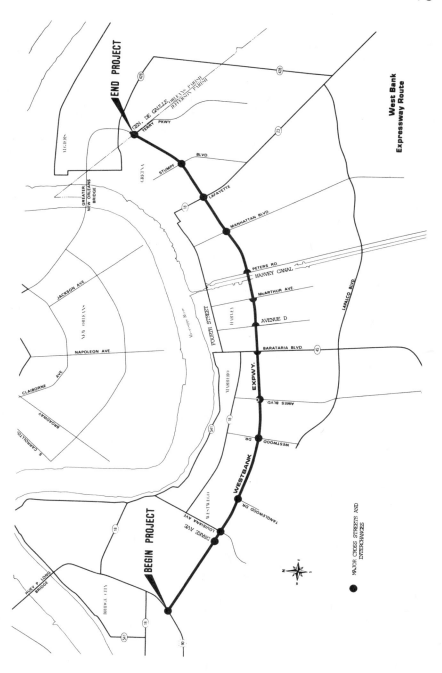

6. *Environmental Effects:*

 a. *Beneficial Effects:* The proposed project will result in a number of beneficial effects:
    (1) reduced traffic congestion within the Westbank corridor, especially at all intersections and at the Harvey Canal crossing;
    (2) provision for projected traffic growth through 1994;
    (3) greater safety for both vehicular and pedestrian traffic along the corridor;
    (4) shorter travel times for both through and local traffic;
    (5) improved ingress and egress provided by the interchanges;
    (6) cross-expressway access at grade separations with considerably improved safety;
    (7) lower noise levels along the elevated portion;
    (8) increased utilization of adjacent lands; and
    (9) better landscaping and aesthetics.

 b. *Adverse Effects:* The proposed action will cause some adverse effects:
    (1) longer travel distances on one-way parallel service roads to reach certain destinations;
    (2) reduction of the number of cross streets from 34 to 22;
    (3) loss of 14 acres of swamp hardwoods from the vicinity of the U.S. Highway 90 interchange;
    (4) reduction of visibility and direct accessibility for some corridor businesses; and
    (5) projected noise levels slightly higher than standards at a few locations along the ground level portion, but less than they would be with continued use of the existing expressway.

7. *Alternates to the Proposed Action:*

Alternates considered were:
    (1) "Do-Nothing"—maintenance of the status quo, with no improvements to the existing expressway.
    (2) Minor Improvements at Intersections—better signalization, land striping, lengthened left turn lanes, sign additions, channelization, and the removal or shielding of hazardous roadside objects.
    (3) Improved At-Grade Expressway—addition of a third traffic lane in each direction, signalization improvements, separate left and right turn lanes, widening and resurfacing of parallel service roads and construction of two new two-lane tunnels under the Harvey Canal.
    (4) Ground-level Upgraded Expressway—identical improvements to the proposed action, except the expressway roadway east of Barataria Boulevard would be at-grade. Grade-separation interchanges would be provided at major cross streets.

(5) Improved Mass Transit—expansion of existing transit operations and addition of flexible feeder bus system.

8. *Comments Solicited From:*

a. *Federal:*

Senator Russell B. Long, Washington, D.C.

Senator J. Bennett Johnston, Washington, D.C.

Congresswoman Lindy Boggs, Washington, D.C.

Advisory Council on Historic Preservation, Washington, D.C.

Department of Agriculture, Soil Conservation Service, Alexandria, Louisiana

U.S. Army Corps of Engineers, Lower Mississippi Division, Vicksburg, Mississippi

U.S. Army Corps of Engineers, New Orleans District, New Orleans, Louisiana

Department of Commerce, Washington, D.C.

Department of Commerce, Economic Development Administration, Austin, Texas

Department of Commerce, National Marine Fisheries Service, St. Petersburg, Florida

Department of Commerce, National Weather Service, Fort Worth, Texas

Community Services Administration, Dallas, Texas

U.S. Environmental Protection Agency, Dallas, Texas

Department of Health, Education and Welfare, Dallas, Texas

Department of Health, Education and Welfare, Public Health Service, Dallas, Texas

Department of Housing and Urban Development, Office of Community and Environmental Standards, Washington, D.C.

Department of Housing and Urban Development, Fort Worth, Texas

Department of Housing and Urban Development, New Orleans, Louisiana

Department of the Interior, Washington, D.C.

Department of the Interior, U.S. Fish and Wildlife Service, Atlanta, Georgia

Department of the Interior, U.S. Geological Survey, Baton Rouge, Louisiana

Department of the Interior, Bureau of Outdoor Recreation, Albuquerque, New Mexico

Department of the Interior, Bureau of Mines, Denver, Colorado

Department of the Interior, National Park Service, Santa Fe, New Mexico

Interstate Commerce Commission, Washington, D.C.

Department of Labor, Manpower Administration, Washington, D.C.

Office of Economic Opportunity, Dallas, Texas

Department of Transportation, Washington, D.C.

Department of Transportation, Federal Aviation Administration, Fort Worth, Texas

Department of Transportation, Federal Highway Administration, Washington, D.C.

Department of Transportation, Federal Highway Administration, Fort Worth, Texas

Department of Transportation, Urban Mass Transportation Administration, Dallas, Texas

Water Resources Council, Washington, D.C.

b. *State:*

Louisiana Commission on Intergovernmental Relations, Baton Rouge, Louisiana

Governor's Council on Environmental Quality, Baton Rouge, Louisiana

Joint Legislative Commission on Environmental Quality, Baton Rouge, Louisiana

Louisiana Attorney General, New Orleans, Louisiana

Louisiana Department of Agriculture, Baton Rouge, Louisiana

Louisiana Air Control Commission, New Orleans, Louisiana

Louisiana Department of Art, Historical and Cultural Preservation, Baton Rouge, Louisiana

Louisiana Civil Defense Agency, Baton Rouge, Louisiana

Louisiana Department of Commerce and Industry, Baton Rouge, Louisiana

Louisiana Department of Conservation, Baton Rouge, Louisiana

Louisiana Forestry Commission, Woodworth, Louisiana

Louisiana Geological Survey, Baton Rouge, Louisiana

Louisiana Health and Social and Rehabilitation Services Administration, New Orleans, Louisiana

Louisiana State Parks and Recreation Commission, Baton Rouge, Louisiana

Louisiana State Planning Office, Baton Rouge, Louisiana

Louisiana Public Service Commission, Baton Rouge, Louisiana

Louisiana Department of Public Works, Baton Rouge, Louisiana

Louisiana Stream Control Commission, Baton Rouge, Louisiana

Louisiana Tourist Development Commission, Baton Rouge, Louisiana

Center for Wetland Resources, Baton Rouge, Louisiana

Louisiana Wild Life and Fisheries Commission, Baton Rouge, Louisiana

c. *Local:*

Regional Planning Commission, New Orleans, Louisiana

Jefferson Parish Council, Gretna, Louisiana

Jefferson Parish Planning Department, Metairie, Louisiana

Metropolitan District Commission, Metairie, Louisiana

Mayor, City of New Orleans, New Orleans, Louisiana

New Orleans City Planning Commission, New Orleans, Louisiana

Mayor, City of Gretna, Gretna, Louisiana
Gretna Planning Commission, Gretna, Louisiana
Mayor, City of Westwego, Westwego, Louisiana
Westwego City Planning Commission, Westwego, Louisiana

d. *Environmental:*

Sierra Club, Delta Chapter, New Orleans, Louisiana
Orleans Audubon Society, New Orleans, Louisiana
Ecology Center of Louisiana, Inc., New Orleans, Louisiana
Louisiana Wildlife Federation, Baton Rouge, Louisiana
Orleans Wildlife Society, New Orleans, Louisiana
League of Women Voters of Jefferson Parish, Marrero,
    Louisiana
American Lung Association of Louisiana, Inc., New Orleans,
    Louisiana
Citizens for Sound Planning & New Orleans Center for
    Housing and Environmental Law, New Orleans, Louisiana
Society for Environmental and Economic Development,
    Gretna, Louisiana
Fund for Animals, Metairie, Louisiana

e. *Others:*

Greater Jefferson Port Commission, Gretna, Louisiana
Jefferson Parish Industrial Development Commission,
    Harvey, Louisiana
West Bank Bridge and Transportation Commission, Gretna,
    Louisiana
Harvey Canal Industrial Association, Gretna, Louisiana
Chamber of Commerce of the New Orleans Area, New
    Orleans, Louisiana

Source: Department of Transportation, U.S. Coast Guard. *Draft Environmental Impact Statement—West Bank Expressway, Jefferson Parish, Louisiana.* New Orleans, Louisiana, July 1976, pp. v–xiii.

The citizen should scan the entire EIS, to gain some perspective on it before engaging in a detailed review.

FIGURE 5   Sample of table of contents for an environmental impact statement.

Draft Environmental Impact Statement
U.S. Coast Guard

West Bank Expressway
Jefferson County, Louisiana

TABLE OF CONTENTS

Source: Department of Transportation, U.S. Coast Guard. *Draft Environmental Impact State-ment—West Bank Expressway, Jefferson Parish, Louisiana.* New Orleans, Louisiana, July 1976, pp. i-iii.

A More Detailed Review

Getting Started. If the group has not already done so, it should contact the agency staff who worked on the EIS and who will receive the comments. They are identified on the title page of the EIS or on the summary page, under the heading "responsible agency." The staff probably will be willing to discuss how they prepared the impact statement and, later on, they may be able to explain any confusing points in it. They also may suggest other agencies or people who can provide additional information.

Establishing a cordial relationship with the staff during the early stages of the review is important, since they can be very helpful at this point. Also, when the time comes to submit comments to the agency on the draft statement, a group's comments may be more carefully considered if there has been previous communication with the agency; the group may be able to avoid becoming involved in the polarization that usually takes place as disagreements arise between the agency and those who do not support the project.

When communicating with agency staff, the group should ask for a copy of the guidelines the agency uses when preparing an EIS. Because these guidelines contain the agency's regulations concerning what should be included in the EIS, they will help the reader spot certain kinds of omissions. And a working knowledge of the guidelines will make it easier for the reader to get through what may be a long and complex document.

Finally, a visit to the project site, particularly if the site is unfamiliar, is usually a good idea. This will not only make it easier for the group to envision the project while reading the EIS, but it also will facilitate meeting some of the people who will be directly affected by the project and may highlight issues not covered in the statement.

Reading the EIS. An EIS should not be read like a book—from start to finish, because the analysis of each environmental impact generally appears in several sections. The most effective way to review a statement is to skip from chapter to chapter. For example, to look in detail at any one impact of a proposed project, the reader must first read the pertinent parts of the "project description" and the "environmental setting." Then he must turn to the section on

Detailed review of the EIS requires focusing one's attention on the arguments and facts that have to do with the particular environmental effects about which one is concerned.

"probable impact" to determine possible impacts, and to the section on "probable adverse impacts which cannot be avoided" to see whether the impact will definitely occur. Finally, the reader must turn to the section on "alternatives to the proposed action" to see whether, if another plan were used, the impact could be avoided.

If a group is available, the major questions about the project should be divided among the members. For example, if a proposed highway is being studied, one member might look into the need for the project, another into the estimates of air pollution, another into whether any major alternative corridors have been missed, and so forth. This facilitates focusing on those aspects of the project which are of the greatest concern and makes each person's task more manageable.

During the review, the pertinent question to keep in mind is "what environmental consequences will result should this action take place, and how can they be avoided?" The objective of assisting agency staff in their review of the proposed project and the objective of influencing the final decision both require that the EIS clearly and accurately disclose the environmental effects of the various choices that are available. Thus, in reviewing the statement, the group must determine whether the conclusions of the EIS are sound and pay particular attention to whether all the major issues have been fully covered.

Are the Analyses Clear and Convincing?

The analysis of each major impact should include a brief description of the key features of the area that will be affected, a thorough explanation of the probable impact, and an indication of the methods used to arrive at the prediction. It also should include an assessment of the significance of the impact. Unfortunately, few impact statements are as clear or thorough as they should be. To give readers a better idea of how to judge a statement, the following sections describe the different elements that it should contain.

Background Information. The agency should provide sufficient background information for the reader to gain an understanding of the context in which the project and, specifically, the impacts will take place. This information might include physical descriptions, statistical information, and descriptions of flora, fauna, and climate, as well as a detailed description of certain aspects of the project itself. There should be some indication of how the agency obtained this information and how reliable it is.

Probable Impacts. The agency should provide a clear and detailed description of what the probable impacts of the action will be and should indicate specifically what the cause of each impact will be. For example, the statement might say that, because of warm water discharge into a lake, one of the impacts of a proposed power plant will be an increase in the overall lake temperature. In reading the statement, the reader should try to judge how *completely* the agency discusses each impact. For example, does the statement explain how the fish will be affected by the increased water temperature in the lake?

Methodology. The agency should attempt to explain how it determined what the impacts of a proposed action will be. If the description of the possible impacts is riddled with words like "might" or "could" or "possibly," the reader may have cause to wonder how the agency arrived at its conclusions. Although it is impossible to be absolutely positive about some predictions, the reader should consider whether the agency appears to have been thorough enough in its investigation.

The general procedure that was used to predict the impacts should at least be explained, and, if the explanation must be incomplete, references to sources with more complete and detailed descriptions should be included.

Significance. Discussing the significance of impacts is difficult, but it is probably the most important part of the analysis. Unfortunately, many EISs are inadequate in this area. The significance of an impact is best shown by placing it in perspective, using the background information already provided.

For example, one EIS reported, without elaboration, that "the project is expected to reduce runoff from 4 to 3½ tons/acre/year." This statement is difficult to evaluate because the reader is not provided with any information concerning normal rates of soil loss. The EIS should have added that a loss of seven tons per acre is close to the minimum under natural conditions in the project area and that losses of up to forty tons per acre are common.

Failing to place impacts in perspective reduces the reader's ability to make judgments on a project. Fortunately, many agency guidelines require that discussions of the significance of impacts be included in EISs, and suggest ways to handle this. Reviewing the agency's guidelines can help readers know what they have a right to expect of the EIS.

Example. Consider the following excerpt from an actual impact statement. It concerns a program to exterminate major blackbird roosts in Kentucky and Tennessee. Since the roosts contained millions of birds, some individuals questioned the effect the program would have on the continental blackbird population. This was a difficult question to answer without recourse to biological jargon, but the authors succeeded in assembling a clear, well-reasoned analysis.

Comments

Impact of Roost Reductions on the Blackbird Populations of Eastern North America

Based on a survey in 1969–70 by biologists of the U.S. Fish and Wildlife Service, the winter roost population of blackbirds in the eastern United States consists of approximately 350 million birds. The population has the following estimated species composition: grackles—25 percent or 90 million birds; redwings—40 percent or 140 million birds; starlings—20 percent or 65 million birds; and cowbirds—15 percent or 50 million birds. Additional species are found in small numbers but are not discussed here. These estimates are only crude approximations and no confidence limits can be given to indicate their accuracy; however, they are the best estimates available.

Here the agency provides background information (the number of birds in the eastern United States) and indicates where it obtained the information.

The methodology used . . .

A crude graphic model is presented in Figure 12 [of the original impact statement] to indicate the general annual cycle of the grackle population in eastern North America. The model assumes that adult grackles have an average survival rate of 52 percent and

and what its
limitations are.

Explanation of why
species other than
grackles are not
included in the model.

Discussion of impact
and its significance.

that adult females annually fledge an average of 2.76 young. These population parameters are typical for passerine species in North America. The model does not pretend to precisely define the numerical response of the grackle population to roost control operations. It is merely an attempt to put into proper perspective the probable immediate impact on the population of eastern North America of a roost reduction of the magnitude planned.

The model is presented only for the grackle population, since this species appears to be predominant at the Fort Campbell and Milan AAP roost sites. The impact of reduction of redwing and cowbird populations would probably be less than on the grackle population, since these two species form only a small part of the roost population.

If we assume the population of grackles in the eastern United States to be 90 million birds in January, and that the roosts at Fort Campbell and Milan AAP contain 2 million and 5 million respectively, a total reduction at these roosts in January would remove about 7.7 percent of the population. This man-caused reduction would be fairly small compared to the natural mortality occurring during the winter months as indicated in Figure 12, and ignores birds in smaller uncensused roosts and those which do not join roosts. Two hypothetical survival curves are presented to indicate the possible numerical responses of the grackle population immediately after such an event. In either case, the proposed reduction would apparently have little effect on the total grackle population of eastern North America.

Contrast the foregoing paragraphs with the ones that follow, taken from a small watershed EIS, which illustrates common EIS errors. No additional information on the points raised in the following passage was included elsewhere in the EIS. This paragraph is from a section on the economic and social effects of the project.

Comment

No background information is given on individual or farm income; there is no *discussion* of the possible economic or social effects; the significance of the effects is not explained.

Economic and Social Effects of the Watershed Program

Per capita income in the watershed will be increased about $95 as a result of increased agricultural income. Average net income of the 100 muckland farms will be increased about $5,000 annually. This will encourage their continued existence and will help to reduce the migration of rural people to urban areas. Increased income of the farm laborers and increases in the tax base will help reduce deficiencies in needed services.

Are There Major Omissions?

After the analyses in the EIS have been evaluated, it is important to consider whether any important issues have been overlooked. Surprisingly obvious omissions sometimes occur in impact statements, and they usually fall into one of the following categories:

Alternatives. It is a mistake to assume that all means of reaching the project goals have been discussed in the EIS. A team at Cornell University reviewed a Department of Transportation EIS concerning a proposed highway that would run through Corning, New York. The highway was expected to ease congestion problems in Corning and complete the Southern Tier Expressway, a major route in southern New York State. One alternative to the project which the EIS considered was making improvements to local streets. This would have decreased congestion, but the plan was rejected because it would not have met the second objective—completion of the Southern Tier Expressway. A second alternative discussed was bypassing Corning with the expressway. This alternative was attractive because it would have avoided the substantial adverse impacts of a major highway through the community. It was rejected because it did not solve the congestion problems. The EIS did not, however, consider the seemingly attractive alternative of an expressway route bypassing Corning and improvements to local streets to ease Corning's traffic congestion problems. This alternative would have solved both the transportation problems and avoided bisecting the community with a four-lane, limited-access highway.

Direct Impacts. Significant environmental effects, on occasion, also are left out of impact statements. In reviewing the EIS covering the proposed corridor for an electric power-plant transmission line, a group discovered that a major agricultural pest—the golden nematode—was not mentioned. The golden nematode is such a serious threat in the project area that farmers try to avoid moving vehicles from one field to another because they may carry the nematode with them to uninfested areas. When machinery must be moved between fields, it is often steam-cleaned to minimize the risk of spreading the pest.

The transmission corridor would be serviced regularly by vehicles traveling from infested to uninfested areas. The nematode's spread, as a result of the corridor's construction, was a serious possible adverse impact which could have caused a multimillion dollar agricultural loss. Yet, it was not mentioned in the EIS.

Indirect Effects. In addition to the immediate, direct impacts of a project, the EIS should discuss the secondary, or indirect, impacts. But because each impact can cause an endless succession of others, it is not always clear where the analysis should stop.

For example, a certain flood prevention program was expected to slightly increase flooding downstream from the project. Obviously, this was an important indirect effect. The flooding would cause crop loss, which could lead to conversion of prime agricultural land to other uses, which could cause farmers to sell their land and move to cities, which could lead to increased unemployment and, perhaps, increased welfare payments. How far should the statement go in its discussion of indirect impacts?

Common sense is a good guide in making this decision. Since most agency investigations are expensive, the possible benefits of an investigation should be worth its cost. Thus, two other

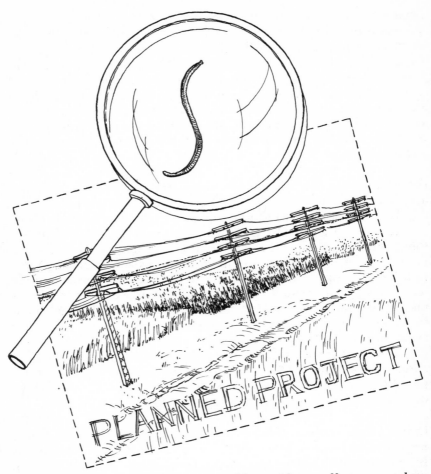

An EIS should consider all the foreseeable significant effects . . . such as the fact that vehicles servicing an electric transmission line may spread the golden nematode, an agricultural pest, from one field to another.

guidelines that should be considered before requesting additional information on indirect effects are:

1. The information should have an important bearing on the design or overall evaluation of the project.
2. The information should be possible to obtain.

Portions of the Description or Analysis. Discussions of some parts of the project, or the environment in which it occurs, also may be left out of an EIS. In the project involving the golden nematode, power corridors were described thoroughly, but consideration of a power-transmitting substation along the corridor was omitted.

Sometimes part of an analysis is left out. Cost-benefit analyses, for example, may include indirect benefits but exclude indirect costs. If a reservoir will be created by a project, the cost-benefit analysis may include the recreation opportunities of the reservoir as benefits. It should then also include, as costs, the loss of whatever present recreation opportunities will no longer be available when the reservoir is built.

The best way to learn of neglected impacts or descriptions, besides thinking about the project, is to talk with people who are familiar with the project and the project area. In addition, experts in the field, who are not familiar with the project, may be able to make useful suggestions. Checking the agency regulations and CEQ guidelines to determine whether each required topic is discussed also may be helpful. An objection, however, should not be made just because a required topic is omitted; there should be a reason for covering it. If there is a topic that the group feels should be addressed, they will help their case if they can show that the regulations or guidelines mandate that it be considered.

Formulating a Position on the Project

Once the citizen or interest group have determined that the draft impact statement is sufficiently accurate and complete, they should use the statement to develop or review their position. They should decide whether the benefits of the project are worth the potential impact it will have on the environment. This usually requires a balancing of "hard values," such as the production of additional electric power or the establishment of a new 100-store shopping center, against "soft values," such as preserving a free-flowing stream or preventing the loss of an endangered species. There is, of course, no formula for making this type of decision. It requires a judgment—one that reflects the particular views of the individual or group.

The members also should decide whether an alternative formulation of the project would make it more acceptable. Is there a way to reduce the environmental damage—for example, through the selection of a different corridor for the highway, or the installation of cooling towers on the power plant—and still preserve some or all of the project's benefits? Again, what is usually involved is a balancing of "hard values" against "soft values." The cost of additional devices to protect the environment or the loss of some of the project's benefits must be weighed against the improvement in environmental quality that will result.

Commenting to the Agency

Before a citizen or group send its comments to the agency, it should try to consider the agency's position. Agency staff are likely to avoid major reconsideration of the project for at least four reasons:

1. They are generally committed to the project. It has probably been in the planning and evaluation stage for years. The project has become the agency's recommendation—its best plan for solving a problem. Moreover, some of the individuals who are responsible for responding to the comments are often the same individuals who have prepared the analyses that others are suggesting should be changed.

2. This is a hectic time for the agency. Numerous, diverse comments are being received, and all must be answered in writing.

3. Agency staff are specialists in solving the type of problem at hand. They may be dubious about the ability of nonprofessionals to tell them how to better carry out their professional responsibilities.

4. There may be other important reasons, not mentioned in the EIS, why an agency desires rapid approval of the project. In one case, the agency that wrote the statement was facing an imminent deadline. Procedural requirements governing the agency's preparation of statements were about to change. If the project was not approved within a few months, the EIS would have to be rewritten entirely and submitted as a new draft EIS.

Because of these pressures, poorly framed comments tempt officials to respond "How can we dispense with these criticisms quickly?" rather than "Is there a point here we should study more carefully?" Consequently, careful consideration should be given to selection of the comments to submit and their preparation for submission to the agency.

Which Comments to Submit. After developing comments, the group should wait a few days and then reevaluate them. It should make sure each request for additional analysis is important, feasible, and not too costly.

If uncertain of the reasonableness of a request, the group should call the agency. This process can help eliminate comments of questionable validity, and it can also help identify the most important points.

Next, the group must decide whether to submit all of its comments. There are three possible approaches:

1. The group might include all comments that have been developed. This strategy, probably the most obvious, has a serious shortcoming. The most significant comments may get lost among the less important ones. If 20 suggestions are made and the agency accepts 18 of them, it will be hard to argue that the agency was unresponsive—even if the 18 comments they responded to were minor flaws and the 2 they ignored were important objections.

2. Another alternative is to submit only a few comments formally, each directed to a major point and extensively supported. This is the opposite extreme. The group may want to maximize agency consideration of a few central questions about which it is concerned. This is a good strategy when the group is convinced that some issue(s) needs extensive consideration before the project proceeds. Focusing on these, and defending the points in great detail, notifies the agency that these issues merit its most careful consideration. The group might then consider submitting its other comments separately, noting that, since agency time is limited, it is not requesting formal replies, although the group hopes the comments will be useful to the agency.

3. Finally, the group might identify the most important comments but submit several others too. This strategy is a compromise between the first two approaches. The major comments get first attention, but some others also get formal consideration. This is a particularly effective approach if the group is generally satisfied with the project and believes the agency is inclined to accept the group's suggestions.

These three strategies suggest the range of alternatives available

Comments on a draft EIS are most effective when they are very selective.

to the group; but, each case must be decided individually. In designing its approach, the group should consider the aforementioned alternatives and recall that agencies have limited time to consider comments and that submitting only important points formally increases the chance that the agency will focus on those aspects of the project that the group considers most crucial.

Whichever strategy the group chooses, if compliments are due the agency on parts of the statement, they should be conveyed together with the criticisms. Finding ways to commend the agency will help win the agency's cooperation and make it easier to maintain cordial relationships throughout the discussions about the comments.

Preparing Comments for Submission. The cardinal rule to follow in preparing comments for submission is "Make them clear." Whenever possible, reference should be made to specific page and paragraph numbers in the EIS.

Agencies are not required to guess what the comments mean.

They must respond to each comment, but vague requests usually receive perfunctory answers. The following are a few examples of poorly framed comments taken from final impact statements.

Comment: Overall project emissions could be reduced by increasing local bus service to the project area.
Response: As the project population increases, more frequent bus service is expected.

Comment: Natural rock formations within the study area should be protected.
Response: Paragraph 4.23 has been revised to incorporate this information.

Comment: Evaluation of flood hazards in compliance with Executive Order 11296 and "Flood Hazard Evaluation Guidelines" should be completed.
Response: The Draft EIS has extensively evaluated flood hazards on the project site. See pgs. 42–44, 115 and 151 of the draft.

The amount of detail and justification the group should provide with its comment depends on the magnitude of the request. If the question is easy to answer, an extensive justification is not necessary. This is illustrated by an exchange which appeared in a small watershed EIS:

Comment: How much of a reduction in wind erosion from the muckland can be expected from this project?
Response: Wind erosion rates will be reduced about 50%.

If the group wishes the agency to carry out additional studies, it should explain why the request is justified. If the group fails to do so convincingly, the agency is likely to reply that it has already studied the issue enough or that it will study it more during the design phase of the project.

Comment: Mixed forest along Stream B should be investigated for use as a linear part or buffer area between the project development and the residential zone area to the east.
Response: Preservation of vegetation is a major objective outlined in the Draft EIS under part 5, Vegetation and Wildlife on pages 149–150, and under part 6, Soils on page 151. The list of controls to be implemented to achieve this objective can be found on page 26 of the Alternatives Section herein.

Comment: The statement should include a detailed analysis of the present land use regulations, state and county and municipal, affecting the area. The question of how the proposed action will affect federal, state and local land use plans and policies is not answered.

Response: A detailed analysis of past land use regulations, State, county, municipal, affecting the area will be included as a part of the development of the land use control measures.

These requests for additional studies did not show why the information was needed to fulfill the EIS requirements. The following comment supported its request more effectively.

Comment: Although the draft statement mentioned . . . the encroachment of the highway on the Normanskill floodplain, it does not include an evaluation of flood hazards in compliance with Executive Order 11296. This Executive Order discussed the evaluation of flood hazards when planning the location of Federally financed or assisted new facilities such as highways. The FHWA has issued memorandum 20-1-67 to implement the Executive Order.

Subsequently, in April 1972, the Water Resources Council issued "Flood Hazards Evaluation Guidelines" which are to be utilized by Federal Executive Agencies in complying with the Order. The final statement should demonstrate that such compliance has been taken.

Response: See page 112 of the Final EIS. [The agency carried out the requested studies, and included them in the body of the final impact statement.]

Agency responses to comments vary in quality. Some agencies make every effort to accommodate requests. Others are less responsive. An examination of two or three final impact statements prepared by the agency will show how the agency has responded to comments in the past.

The Overall View on the Project. The citizen or group should not forget to state whether it is, overall, in favor of or opposed to the project. In either case, it should indicate why. The more specific the group can be, the more weight its comments are likely to carry. For example, if the group is opposed to the project, an alternative may be developed which will preserve some or all of the benefits and which will avoid the specific problems identified in the group's comments. Objections must be clearly presented so an agency can know where to begin to try to overcome them.

Occasionally, an impact statement is so defective that an individual or group is unable to decide whether the benefits of the project outweigh the costs. There may simply not be sufficient information to reach a conclusion. In such a case, the citizen or group should urge the agency to prepare a new draft impact statement and circulate it for comment. The agency should be advised that the draft statement lacks so much information that an overall assessment of the project is impossible.

5

The Final
Impact Statement
and the Decision

This chapter covers the concluding steps in the environmental impact statement process: the issuance of the agency's final statement and the decision on whether to proceed with the project. It describes how the agency writes the statement, discusses ways to review it, and, finally, offers suggestions on how to influence the agency's decision on the project.

Final Environmental Impact Statements

In most cases, the final environmental impact statement is a composite of the agency's draft environmental impact statement, copies of all the comments on the statement, the agency's responses to these comments, and revisions to the body of the statement itself, incorporating some of the criticisms or suggestions received. The comments and the agency's responses often are found in an appendix at the end of the statement.

If the agency becomes aware of additional information on the project as a result of its own investigations, this new information may be included in the final statement. The agency also may use the final statement to clarify or expand on points made in the draft. No major additional environmental consequences or alternatives, however, should appear for the first time in the final statement, regardless of whether they came to light as a result of the agency's own analysis or from comments the agency received on the draft

The lead agency reads citizens' comments on the draft EIS. Both the comments and the response are included in the final EIS.

statement. If such new information exists, it means that the draft statement was defective—it didn't fully delineate the environmental impacts and reasonable alternatives to the proposed action. If this new information substantially changes the picture presented in the draft statement, the agency must issue and circulate for comment a revised draft or supplemental statement before issuing a final statement.

Basically, under NEPA, the issuance of the final statement ends the official role of the public in the environmental impact statement process. This does not mean that there is nothing more citizens can do at this point; in fact, there is a great deal that can be done. But there is no longer a formal, established channel through which citizens can communicate with the agency.

The release of the final impact statement signals the beginning of the last stage in the impact statement process. According to the CEQ guidelines, an agency must allow 30 days for its officials to review the final statement, after which the agency may, at any time, make a final decision on the proposed action. In actual practice, if the proposed action has progressed this far in the review process, the agency will in all likelihood approve it. Thus, for the concerned citizen who is still unhappy with the agency's point of view, this 30-day waiting period is crucial. It represents the last chance a citizen has to influence the agency's decision. Some suggestions on how to influence the agency's decision are given later in this chapter.

Before the final EIS is released may be an opportune time to mobilize other agencies, speak to civic groups, and contact members of Congress, state legislators, and members of the city council.

When to Expect the Final Statement

Most agencies issue the final statement within one year after release of the draft. The average time between the release of the draft and final statements, however, varies substantially from agency to agency. For example, in 1974, the average elapsed time was 3½ months for the Bureau of Land Management, 5 months for the Department of Housing and Urban Development, 9 months for the Federal Aviation Administration, 12 months for the Federal Highway Administration, and almost 20 months for statements written by the National Park Service. Table 3 presents the average length of time between the issuance of draft and final statements for each of the major federal agencies. In cases where the proposed government decision is not highly controversial, much of this time is spent writing the final EIS. If the draft EIS was not well prepared, or if the project is particularly complex, release of the final state-

TABLE 3 Average Time Between Filing of Draft and Final EISs, Calendar Year 1974 (in months)

Agriculture	
Forest Service	9.1
Soil Conservation Service	12.7
Commerce	4.4
Defense	NA
Air Force	
Army	
Corps of Engineers	10.5
Navy	
Health, Education, and Welfare	NA
Housing and Urban Development	5.2
Interior	NA
Bureau of Indian Affairs	3.6
Bureau of Land Management	5.7
Bureau of Outdoor Recreation	6.1
Bureau of Reclamation	11.7
Fish and Wildlife Service	19.8
National Park Service	13.0
Geological Survey	6.7
Justice	
Law Enforcement Assistance Administration	4.0
Labor	2.5
State	14.4
Transportation	
Federal Aviation Administration	8.9
Federal Highway Administration	11.9
Treasury	NA
Energy Research and Development Administration	6.1
Environmental Protection Agency	7.4
Federal Energy Administration	NA
Federal Power Commission	5.3
General Services Administration	6.5
Nuclear Regulatory Commission	NA

NA = Not available

Source: Council on Environmental Quality. *Environmental Impact Statements: An Analysis of Six Years' Experience by Seventy Federal Agencies.* Washington, D.C., U.S. Government Printing Office, 1976, p. 30.

ment will take longer than usual. At least two and one-half months are necessary, however, since agencies usually allow one and one-half months for the comment period on the draft, and at least one month is required for writing and processing the final version through the agency's administrative channels.

In cases where the proposed action is highly controversial, the agency may spend much of its time attempting to resolve the controversy before the final statement is released. Many agencies use this period to work out differences of opinion that surfaced in comments submitted on the draft. During this time, the agency also will be performing additional analyses and meeting with other agencies. Finally, the agency will use this time to alter the project to make it more acceptable to its critics, if the agency feels criticisms received were well founded.

In some cases, agencies will not file a final impact statement. If the comments on the draft indicate such widespread opposition that approval is highly unlikely, the agency may abort the impact statement process and cancel further consideration of the project. At the other end of the spectrum, if few or no comments are received on the draft statement, the agency may conclude that an EIS was not in fact required. It may attempt to justify this conclusion on the ground that, since no comments were received, the environmental consequences probably are not significant; the agency will then proceed to make a decision on the project without a final statement.

Getting a Copy of the Final Statement. Agencies do not widely circulate final statements and may not even publicize their existence. The best way to get a copy of the final statement is to ask for one when submitting comments on the draft. Those who did not submit comments on the draft should ask that their name be put on the mailing list for copies of the final statement.

After release of the final statement, a copy can be obtained by calling the involved agency. (Writing will take too long, since the agency may wait only 30 days to make its decision on the action.)

As with draft statements, agencies generally distribute final statements free of charge while their supply of copies lasts. In a few cases, a small charge is made. The only exception is when the statement is unusually long—the Trans-Alaska Pipeline statement, for example, filled six volumes, with a three-volume appendix. While most of their statements are free, in this case the Department of the Interior charged approximately $25 per copy.

In the case of a power plant, the agency may meet with an executive from the utility, before the final statement is issued, to find ways to reduce further the project's environmental damage. A meeting is more likely to occur if numerous critical comments are received on the draft statement, because it signals to the agency that the environmental consequences of the proposed project are unacceptable to many citizens.

Reviewing the Final Statement

Objectives. The citizen should have two objectives when reviewing a final impact statement. First, he should reevaluate his position on the project. If he submitted comments on the draft, he can readily do this by reading the agency's response to his questions and suggestions.

Second, the citizen should try to find political support for his position by identifying others who share it. This, of course, is done by reading the comments on the draft submitted by others. Reading the comments of others also may identify issues that were initially overlooked.

Reviewing the Statement. The first step for those who submitted comments on the draft statement is to proceed directly to the "Comments" section of the final statement, which should contain the text of all letters received by the agency—unless there were so many letters and they were so long that the agency summarized them. The reader should check to see whether his comments were included and, if not, should contact the agency immediately to find out why.

Next the reader should try to determine the adequacy of the agency's response to his comments. Did the agency understand the comments? Did it directly answer any questions? Is the response directed specifically at the comments received? While these questions may seem trivial, it is important to remember that agencies have been known to "misinterpret" questions or to respond in ways that talk over, under, and around a question without ever actually answering it.

An agency's response to comments may be inadequate for a number of reasons: the comment was received but elicited no response; the response was perfunctory, that is, "Comment Noted" with no discussion or changes in the text of the impact statement; the response was not to the point, or merely repeated the original explanation contained in the draft; the agency gave no documentation (data, references) to substantiate its arguments; or a point is conceded and it is major, but there is no change in the agency's position.

If the response is inadequate, the citizen should contact the agency and request an explanation. He should indicate why the agency's answer does not seem responsive and ask for additional clarification of the agency's views on the question. If the comment raises a major issue, the impact statement may be declared inadequate by a court, and the agency may be ordered to rewrite the statement before proceeding with its decision on the project.

The reader may find that the agency's response is adequate even though it does not change his position on the project. He should not consider the only adequate response to be one that concedes his points; rather, an adequate response is a good faith effort on the part of the agency to understand the comment and to respond in a manner that is justifiable in light of the environmental consequences. Adequate responses would include those that fully answer questions or present logical counter arguments, as well as those that concede a point and indicate that the action will be modified.

The next step for readers who submitted comments—and the first step for those who did not—is to look at the comments that were submitted by other federal agencies and groups and the agency's response to them. Again, this is done for two reasons: to find allies for a position, and to learn of new issues. The section headed "Summary and Discussion of Comments," if there is one, will help identify additional issues.

Check the Comments of Other Agencies. Most federal agencies, when commenting on another agency's proposed project, shy away from becoming an advocate for or against the project. They want to help uncover all the major environmental consequences; but, because of limited resources, a reluctance to become involved in a controversy, or a lack of support within the agency itself, they may refrain from following up on their comments on the draft statement.

Comments on the draft EIS are published at the end of the final EIS. By reading the comments of other groups, one may discover potential allies.

The citizen, then, should carefully examine the comments submitted by other federal agencies and, if the points raised are appropriate, be prepared to make them a part of his case. Sometimes an agency can be persuaded to follow up on its comments if the lead agency failed to respond to them. More often, however, citizens must take up the agency's point.

Reviewing the Position on the Project. When the reader has studied all of the important information in the final impact statement, including the comments and the agency's responses, he should review his position on the project. It is important to remember, however, that the agency can make its decision in as few as 30 days from the date of the release of the final statement. If a citizen wishes to influence the agency's decision, he must work quickly.

Influencing the Government Decision on the Project

Such Action May Not Be Necessary. The citizen may find, for a number of reasons, that he agrees with the government's decision on the proposed action. The agency may have made changes in the project to eliminate certain environmental risks, or it might have shown that the environmental consequences will be less serious than the critics had suggested. While the agency is unlikely to indicate in the final statement that it will not proceed with the action (in which case the final statement probably would not have been written), there is the possibility that the government might present a sound and well-reasoned argument that the action will be beneficial.

If the citizen agrees, then his job is almost over. He might want to send a letter to the agency endorsing the project and, if he has worked with agency staff, thanking them for their cooperation. The only other thing the citizen might do is check back with the agency to make sure that the actual decision does not differ from the one suggested by the statement. This is best done by periodically checking with middle-level agency staff—one more reason why establishing friendly relationships with agency personnel is helpful.

Since final agency decisions rarely differ from proposed decisions, things should flow smoothly from this point on. There have been times, however, when pressures from within the agency or from outside sources have caused agencies to make a decision that was not indicated by the EIS. For example, under Secretary Coleman, the Department of Transportation's decisions generally reflected the opinions expressed in its EISs. Secretary Coleman, however, allowed limited entry into the United States of Air France's SST, even though the EIS did not favor this decision. The citizen, then, should be prepared to act if the agency's decision appears likely to differ from the one suggested in the EIS.

If the Citizen Objects. If the citizen objects to the decision that is indicated in the final EIS, there is just one course of action that remains: mounting a campaign to stop the project.

Taking Stock

It is important to reemphasize here that there is no clear framework for citizen action at this point. The environmental impact statement process is basically over. The agency has prepared a draft environmental impact statement, circulated it for comment, and revised it to reflect the comments received. The National Environmental Policy Act assumes that the agency will use the final impact statement in its decision making and will make a reasonable decision in light of the disclosures in the impact statement. NEPA does not require that an agency select the most environmentally beneficial alternative, but only that it *understand* the environmental consequences of its actions and *consider* them in its decision making. An agency may proceed with an action that involves environmental damage if it is convinced that there are economic and technical benefits that override the environmental drawbacks.

This does not mean that an agency acts without constraint. While it may choose from among different alternatives, an agency may not, under the Administrative Procedures Act, select an alternative that is "arbitrary" or "capricious"—one which no "reasonable man" would select. Thus, although under NEPA an agency is not required to choose any specific alternative, an agency may not make a decision that it alone thinks is valid.

The skeptical reader may well question the assumption that the agency will make a reasonable decision in light of the disclosures in the impact statement. Indeed, it is clear that this does not always happen. In early 1977, a U.S. District Court judge handed down an important decision in *County of Suffolk v. Secretary of the Interior*.[1] The case concerned the Secretary of the Interior's plan to lease land off the Atlantic coast for oil and gas development. The Department of the Interior had gone through the environmental impact statement process, and the secretary subsequently announced that he planned to issue permits. Suffolk County, together with a number of other plaintiffs, sued to prevent the issuance of leases.

Suffolk County argued that the Department of the Interior approached the environmental impact statement process not as a method for uncovering the environmental consequences of a pro-

---

[1] *County of Suffolk v. Secretary of the Interior*, Nos. 75 C 208 and 76 C 1229 (E.D.N.Y. Feb. 1977), *rev'd*, 562 F.2d 1368 (2d Cir. 1977), *cert. denied*, 46 U.S.L.W. 3518 (1978).

posed action, but as an administrative hurdle that the department was forced to surmount. The county argued, in effect, that the Secretary of the Interior had already made up his mind on the action and complied with the impact statement process only to meet the requirements of NEPA—that the environmental impact statement process was, in this case, only a formality. The court found substantial merit to these arguments. Judge Weinstein noted in his decision that there was abundant evidence to conclude that the decision to issue leases was reached before the agency went through the EIS process. The judge concluded, however, that he need not answer the question definitively, because it was clear that the inadequacy of the impact statement in itself would be grounds for overruling the decision. Accordingly, he set aside the agency's decision and ordered the Secretary of the Interior to prepare another impact statement.

The decision in this case was subsequently reversed by the U.S. Court of Appeals, which ruled that the statement was adequate. The appellate court also said that Judge Weinstein should have limited his review to a consideration of the completeness of the impact statement and not concerned himself with the agency's decision-making process. Nonetheless, Judge Weinstein's point—that agencies sometimes fail to use impact statements in their decision making—is well taken, even if he was later overruled on the grounds that the power of the courts is limited in this situation.

Although, formally, the environmental impact statement process may be over after issuance of the final statement, it is important to remember that there can be substantial variation between the theory and the practice of how the impact statement will be used. If a citizen has involved himself in the impact statement process from the beginning, however, he will have substantial tools (agency contacts, familiarity with the project, and a document that points out the environmental consequences of the proposed action) to mount a campaign to fight an agency decision which he feels is irresponsible.

Building the Case

Why Does the Citizen Oppose the Project? Before the concerned citizen tries to persuade others to oppose the project, he should make sure that his own reasons for opposition are sound and, further, that they address specific issues. For example, a citizen who opposes a nuclear power plant because he feels a vague uneasiness about nuclear power is not likely to be very successful in mounting a campaign to stop the project. On the other hand, a citizen who opposes a particular project because it is on an unstable site, the reliability of the safety controls has not been proven, and only a temporary solution to the radioactive waste-disposal problem has been found might be far more successful.

Reasons for opposition. There are a variety of procedural and substantive reasons why a citizen might be opposed to the project at this stage. Organizing the issues of a particular case into one or more of the following categories will help the citizen present a clear argument.

Procedural: (a) The lead agency did not follow its own or CEQ's guidelines in the preparation and review of the EIS. For example, it did not hold required public hearings, did not widely circulate the draft, or did not follow its own guidelines as to what should be included in the EIS.

Substantive: (b) The environmental analysis in the final statement is still insufficient as a guide to sound decision making.

              (c) The expected impacts are severe enough to require modification of the project, implementation of one of its alternatives, or termination of the entire proposal.

              (d) The stated purpose of the project is invalid. For example, a highway is not really needed.

Generally, it is too late at this stage to comment on the procedure followed by the agency (point a) or the adequacy of the statement (point b) unless there are such glaring inadequacies that sound decision making cannot proceed. If this is the case, the citizen should consult a lawyer about a possible lawsuit. Many projects have been delayed by a court injunction because the agency did not follow the proper procedures under NEPA or the impact statement was defective. The citizen might be able to pressure the agency into postponing the decision pending further work on the statement or the fulfillment of certain requirements.

The citizen who is determined to *stop* an action eventually will have to focus on issues that fall under points c and d. In most cases, the project will be halted by convincing the agency, with the help of others, of these points.

The citizen should organize his case in such a way that it is clear not only to himself but also to others, since his major task at this stage is finding support for his point of view. Credibility is critical; therefore, he should be sure to use facts that can be proved. He also should be careful not to distort the issues and to choose a judicious way of presenting his position, because prospective supporters are likely to be disillusioned or wary of becoming involved if they feel he is not being straightforward or is making legitimate points in an exaggerated or highly emotional style.

After the citizen has carefully outlined his own arguments, he should devote some time to identifying the major arguments of his opponents. In order to convince people, it will be necessary to refute opposing viewpoints; thus, it is important to understand the reasoning of those who support the project. This process can be started by examining the "Comments" section in the impact statement. After identifying his opponents' arguments, the citizen should develop counter arguments. These counter arguments should be as well constructed and documented as were the citizen's original arguments against the action and should answer the specific points made by opponents. If the citizen finds he is unable to do this in one or two cases, he might want to consider modifying his position.

Finally, the citizen should try to identify the views of all the major parties involved in the impact statement process—various officials in the agency (they may not all agree), other agencies (federal, state, and local), and interest groups (environmental groups, civic groups, business organizations)—in order to assess potential support and opposition. Again, the "Comments" section of the final impact statement should reveal the positions of the various groups.

The citizen should be careful not to be too hasty in categorizing groups or persons as opponents to his point of view. In a recent case involving lobbying before Congress on amendments to the Clean Air Act, a firm that supplies equipment to auto manufacturers aligned itself with environmental groups. Congress was considering whether to delay the imposition of a requirement that catalysts be installed on new cars. Environmental groups wanted to see catalysts installed, and the equipment supplier, even though it had

Public meetings can provide citizens with an opportunity to present a different point of view on the project, to illuminate defects in the agency's planning, and to generate widespread support. Such meetings can be an extremely useful part of the organized campaign to stop a project.

a close relationship with the major auto companies, made a substantial contribution to the environmental lobby because it manufactured catalysts. The firm felt that, in this case, its interests lay with the environmentalists, not with the auto industry. The citizen should consider the possibility of finding allies in unlikely places.

Pinpointing the Decision Makers

Once the citizen has outlined his position, organized his arguments, and started to line up support, he must consider his overall strategy. It is impossible to prescribe one strategy that applies to all situations, but there are a number of steps the citizen will need to take in most cases. The first is to find out specifically who will make the agency decision.

An examination of the agency guidelines covering the type of action in question will make it possible to determine who, by law, has the authority to make the decision. The citizen should forget

At the conclusion of the impact statement process, the federal agency must make a decision on the proposed project.

about influencing middle-level staff at this point and concentrate on the regional administrator, assistant secretary, or secretary who will be making the decision.

Investigation may reveal that the person empowered by law to make the decision tends to rubber-stamp recommendations made to him, in which case, identifying the person who will make the recommendation is essential. In other cases, it may be clear that the decision maker has already made up his mind; therefore, the citizen must try to reach someone with more authority, who might be willing to review the impending decision.

The Agency's Procedures. After determining who the decision maker will be, the citizen should acquaint himself with the remaining steps in the agency's decision-making process. This may require some digging, because actual customs may differ from the formal procedures. It is extremely important that the citizen ascertain how an agency functions, in order to pinpoint when, how, and where the decision will probably be made.

Sometimes it is possible to determine the criteria that govern the agency's decision. In some cases, these criteria are established by statute or in the agency's regulations.

Someone who has previously worked with the involved agency would be of great help here, and the citizen may want to contact different groups to find someone with the appropriate experience.

Tactics

Once the citizen has determined who the decision maker will be, what procedure the agency will follow, what criteria the decision maker will use, and who supports and opposes the project, he is ready to begin his attempt to influence the decision. At this point, the citizen might contact others who have tried to influence this particular agency's decisions in the past, to find out which strategies worked. The citizen has numerous alternatives here. It is important that he choose his tactics carefully, especially since his time and resources will be limited.

Only a few of the possible options will be described here. Appendix F lists a number of books that focus specifically on strategies for influencing government decisions.

One way to put pressure on the agency decision maker is through direct opposition to his proposed decision by individuals and institutions. Accordingly, the citizen may want to go to other federal agencies, which he has identified from the "Comments" section as opposing the project, and ask them to protest the agency's decision. The citizen can begin by approaching the agency staff who wrote the comments and showing them how the lead agency has ignored their comments in the final impact statement. He should urge them to write to the lead agency and express opposition to the impending decision, and point out that it is their responsibility to do so.

One agency, EPA, has a statutory responsibility to refer to the President's Council on Environmental Quality any proposed project that it determines is unsatisfactory from the standpoint of health, welfare, or environmental quality. If this is the case with the specific project about which a citizen is concerned, he should urge EPA to publicly oppose the project and to refer the matter to the council. In the summer of 1977, CEQ issued a memorandum expanding this arrangement with EPA to include *all* federal agencies. The council asked each agency to refer to CEQ, on a continuing basis, any project that it judges to be environmentally unacceptable. Thus far, approximately six projects have been referred to CEQ, and, while all the results are not in, it appears that the council will be successful in getting many of them modified or halted.

Citizens should not be deterred by an initial hesitation on the part of EPA, CEQ, or any other federal agency to intervene. Agencies shy away from controversy; besides, they have a lot of other work to do. Accordingly, citizens may have to exert pressure to get an agency to intervene, even though intervention is in the public's overall best interest. One way of exerting this pressure is to make clear to the agency that it may be called on publicly to intervene and that the lead agency's failure to consider its comments and views on the project will be publicized. Agencies do not like to be embarrassed, and this may be sufficient inducement to get them to intervene. In many cases, it is *not* sufficient to get the middle-level agency staff to agree to send a letter protesting the proposed decision. The regional administrator or a senior official in the agency must endorse going on record and publicly opposing the proposed decision.

Another way to put pressure on the lead agency is by enlisting the support of a state or local agency. Citizens also should try to bring elected officials into the controversy. By definition, the proposed project will significantly affect the environment and, thus, should concern legislators at the local, state, and federal levels.

Legislators can intervene in two ways. First, they can write to the agency, express their interest in the project, and ask that the agency answer their questions. The citizen should provide the legislator with an outline of the important questions and issues, and try to convince him to request a meeting with agency officials and to write a letter opposing the action. This letter should be sent to the secretary of the agency, with copies to other agency officials, CEQ, and the media.

The second way a legislator can intervene is by sponsoring a resolution opposing the project. The formal proposal of such a resolution often will have a major effect. If the resolution is passed at the federal level, the project will be stopped, since the agency would be violating the law if it proceeded. In most cases, a federal agency is not legally prohibited from going ahead with a project that is opposed by a resolution passed by the state legislature or a city council, but such indication of substantial opposition to the

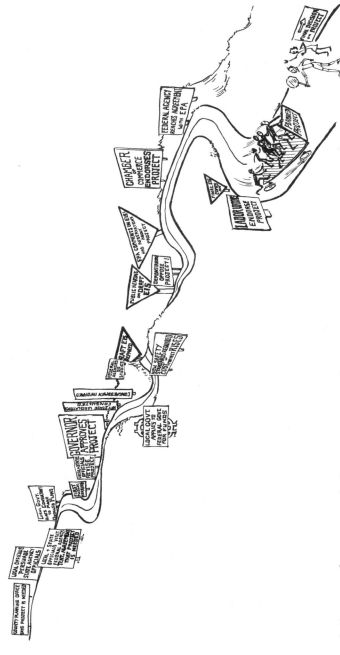

As more and more people become committed to a project, it gains momentum. It becomes difficult to change the plans or to stop the project.

proposed action would likely make the federal decision maker hesitant to approve the project until the objections were overcome.

Another element of any effective campaign is selective use of the media. An effective media campaign must be tailored to fit the needs of a particular situation; in some cases, high visibility is desirable, in others it might not be. Such a campaign might include any or all of the following:

1. *Letters to Newspapers:* Letters to the editor can be very effective, because they educate the public about the proposed action and point out that a controversy exists. Letters are especially effective if several of them are written to the same newspaper by different individuals and groups.

2. *Public Service Announcements:* Radio announcements and editorial comments on television inform and influence many people. Announcements should be straightforward, factual, and comprehensible to the average listener.

3. *Press Releases:* Press releases that are written in a journalistic style—that is, concisely worded, to the point, factual, and objective in tone—and are newsworthy in content are the ones that are most likely to be printed. Press releases are useful for keeping the public current with important developments.

4. *Media Events:* Events such as demonstrations, debates, and presentations of petitions to government officials all focus attention on the controversy. But they should be handled carefully, because they reflect the group's image. The press should be notified in advance of these events.

5. *Press Conferences:* These can be helpful, but only when there is something important to announce.

The organization of support for a media campaign is important. One person cannot handle all the work involved; a number of people must be available to write letters, attend meetings, or demonstrate on short notice.

Citizens who are inexperienced in public relations should not be discouraged if they are not immediately successful in gaining media attention. Talking to media people can provide ideas on how to mount a more effective campaign.

Finally, citizens should not forget the possibility of legal action. Although they are usually expensive and time-consuming, lawsuits have held up many projects. There are three types of potential lawsuits under NEPA. First, a lawsuit may delay a project when an environmental impact statement is required and the agency does not prepare one. Second, a lawsuit may delay a project when an environmental impact statement has been prepared but is inadequate, either because it does not cover all the environmental consequences and the major alternatives or because it does not comply with any of a number of other legal requirements, such as presenting all major points of view in the body of the statement. Third, a project may be halted because the decision is "arbitrary and capricious" in light of the disclosures in the impact statement. Citizens should consult a lawyer to help them decide whether a lawsuit should be filed in a particular case.

Developing the Strategy

A carefully planned strategy is critical at this point; citizens should not simply mount their opposition with only a first step in mind— with no overall blueprint. A team should be organized to plan this effort.

One thing that must be given early consideration is funding. Some members of the group will likely have to spend time seeking contributions either from the general public or from a few individuals who might be willing to offer more substantial support. This is especially important if legal action is contemplated as part of the strategy. While a lawsuit may cost as little as several thousand dollars—if the issues are simple and the lawyer's time is free—a more typical lawsuit is likely to cost upwards of $10,000, and major lawsuits run between $25,000 and $100,000.

Because any campaign may take several months or more, the citizen or group should set some interim goals. For example, while the goal of a campaign might be to stop a project, an interim goal might be to have the regional administrator of EPA voice his opposition. Such goals serve two purposes. First, they provide encouragement for those working on the project. Second, they provide some tangible way of assessing progress and help groups gauge the degree of success of different strategies.

Unforeseen events, however, are bound to occur along the way. In one case involving a school, individuals in the neighborhood opposed the town school board's proposal to close the school. They appeared at a public hearing and announced their opposition, but were unsuccessful in preventing the board from ordering the closing of the school. Immediately upon the board's decision, the individuals filed an appeal with the state department of education. They put together a brief with only limited use of a lawyer, who advised them on how to carry out the appeal. During this period, they also mounted a substantial letter-writing campaign. Several months after the appeal was filed, the state agency declined to overturn the school board's decision. In the meantime, however, so much pressure had been put on the local agency members, and the defects in their original reasoning had been so widely publicized, that two of the board members who had supported the decision decided to resign. Two new members were added to replace them, and with the new composition of the board, there were enough votes to overturn the earlier decision. Thus, while the appeal was unsuccessful in one sense, it turned out to be very successful, indirectly, by challenging the legitimacy of the decision, leading to the resignation of two members, and achieving the reversal of the original decision.

Figure 6 presents the strategy developed by the Southwest Research and Information Center to stop the planned disposal of nuclear wastes in New Mexico. In this case, the NEPA process is just one of several approaches being pursued by the center. The strategy paper briefly outlines the politics surrounding the decision and the manner in which technical evidence, organizing, social research, media, and the law will be used. Note how the various tactics discussed in the preceding section have been integrated here into an overall blueprint.[2] While the paper describes a much larger campaign than will be necessary or possible in most cases, it does show the careful planning and coordination that is always required if a campaign is to succeed.

Final Decision on the Project. A well-organized small group can exert a substantial—indeed a surprising—amount of leverage on an agency. It may make the essential difference between an agency decision going one way or the other. Citizens, however, should not expect to win every controversy they get involved in. While the agency may attempt to put together a proposed project that pleases the largest number of people, it may not be successful, and in any case, not everyone will be pleased with every decision.

---

[2] The only topic that has been deleted from the strategy paper is funding. An attachment to Figure 6 contained a detailed 15-month budget and a discussion of planned fundraising activities, including approaching wealthy donors, arranging benefit concerts and movies, direct-mail solicitation, and proposals to foundations.

FIGURE 6   Strategy paper developed by a public interest organization for an environmental controversy.

Southwest Research and Information Center
P.O. Box 4524   Albuquerque, New Mexico 87106

February 20, 1978

PROGRESS REPORT ON SOUTHWEST RESEARCH AND
INFORMATION CENTER'S CAMPAIGN TO STOP
NUCLEAR WASTE DISPOSAL IN NEW MEXICO

Southwest Research and Information Center is fighting the Department of Energy's plan to dump all nuclear waste in New Mexico. DOE's plan, the Waste Isolation Pilot Plan, (WIPP), calls for storing radioactive waste in salt formations in southern New Mexico near Carlsbad. Salt disposal is clearly questionable. Even those who originally supported the plan now agree that there are many unanswered questions. Strong scientific data developed by Charles Hyder of Southwest Research clearly indicates that salt is not stable in the presence of radioactive waste cannisters and is therefore undesirable as a storage medium.

Waste disposal, according to industry and government sources, appears to be one Achilles Heel of nuclear power. For over 30 years the government has tried a number of methods for disposing of the waste, none of which has been successful. Desperate to find a quick solution, DOE has turned to New Mexico, a state where there appeared to be strong pro-nuclear support. New Mexico prides itself as the birthplace of the atom bomb. Uranium is mined here, basic research continues at Los Alamos and Sandia Labs. As a major employer, the industry seemed to have tremendous business and public support.

It is fair to say that waste disposal is an issue of great immediate concern to New Mexicans. Despite industry's efforts, anti-WIPP groups are springing up in various parts of the state. These groups look to Southwest Research for information, overall coordination, technical assistance and guidance.

For our part, Southwest Research is underway with the campaign. We are developing overall strategy, reaching out to local anti-WIPP organizations, developing a media presence, starting our social research effots, presenting our technical arguments to citizens and the scientific community, hiring staff, undertaking legal actions, and have found a suitable campaign headquarters. We are proceeding to develop the campaign as outlined in the initial proposal which we believe will defeat WIPP.

This update will detail our progress to this point, plans for the future, and our funding situation.

UPDATE

1. Campaign Staffing

Katherine Montague is working as project coordinator, John Liebendorfer as one of the organizers and Ken Schultz as half-time office manager. We have hired a secretary and are currently looking for two community/electoral campaign organizers, a researcher and a half-time lawyer. Lou Colombo, our social re-

FIGURE 6 (continued)

searcher, has moved to Albuquerque and joined the project. Jeff Nathanson,
the media director, is arriving in Albuquerque in mid-March to join the campaign. He has made arrangements with an L.A. media group called Loudspeaker
to provide technical assistance in our media production work. Loudspeaker has
previously done the media for numerous political and referendum campaigns like
the Farmworkers' Prop. 14 in California..

2. Campaign Headquarters

We have rented office space for the campaign — a very business-like building
at 1824 Lomas, N.E., Albuquerque. We move there in 10 days. There is plenty
of space for 12 or 14 people, a meeting area, and it is a good solid building.

3. Overall Coordination

While campaign work was going on in New Mexico, Jeff and Lou spent about six
weeks contacting individuals and groups responsible for organizing anti-nuclear
campaigns on the west coast. They also spoke with a wide range of community
organizers and campaign organizers. In mid-January, Jeff and Lou met with the
Southwest Research staff in Albuquerque to pool information and set overall
campaign strategy, establish a general time line for this year and a detailed
time line for the next two months, and to delineate campaign staff responsibility.

4. Politics

The political issues regarding WIPP are exceptionally volatile. The major candidates for state-wide office, and virtually all candidates for state and national
office are either treading lightly or are on record as opposing radwaste dumping
in the state.

As seen from the attached newspaper articles, WIPP has become a major issue.
DOE has been less than candid and has been caught misleading the people of New
Mexico. Both the State Legislature and the New Mexico Environmental Improvement Agency have formed study panels in the last month to evaluate the effects
of WIPP. To some extent we see these as government efforts to defer responsibility in reaching a decision on the issue, to deflect political pressure and to
counteract negative publicity.

DOE has set up a series of public meetings around the state to explain the WIPP
project. Although the rules havenot been set down, we are working with local
groups to open the meetings up for questions and possible public comment. Most
important, DOE has been forced to concede that the people of a state have veto
power over waste disposal such as the WIPP project — and that it is up to the
state legislature to decide how the people are to express their decision. All
of these actions seem to be responses to recent political pressures.

Thus, the political environment is conducive to the flow of anti-WIPP information
to the public. The study panels and public meetings represent new forums in
which we can make our technical case. The electioneering for federal and state
office gives us abundant opportunities to provide the public and candidates
with needed information. Additionally, the attention given to the issue makes
our organizing work proceed more easily.

One limitation of the current funding situation is lack of money to produce and
distribute media pieces such as radio and TV public service announcements, information brochures, newsletters, direct mail pieces, etc. in the next months.
We feel it is critical to have the funds to reach the mass audience and the
media gives us access to this audience. Hence the absence of adequate funding
is a real impending limitation to our campaign.

5. Important Recent Changes in the Political Environment of the Anti-WIPP Campaign

The state legislature last week rejected a bill that would have forced a state-
wide referendum on a constitutional amendment to prohibit import of radwaste to

FIGURE 6 (continued)

New Mexico. The proposed amendment would have changed our whole campaign into a "confrontation at the polls" in November. We felt that while we may want this confrontation eventually, we needed more time to organize for such a campaign. A year from November we will be much better prepared for such a fight, should we fail to kill WIPP in some other way before that. (Note: The Bureau of Business Research at the University of New Mexico found in January 1977 that a random sample of New Mexicans opposed waste dumping in the state, while favoring the general development of the nuclear industry. Although these attitudes are certainly subject to industry media manipulation, at this point they are promising.)

We were, of course, concerned about the amendment and were in touch with those legislators and individuals supporting the amendment. We saw the bill as a way to make the issue more salient to New Mexicans, to show individuals and organizations nation-wide that there was opposition in New Mexico to waste disposal, and to draw out state legislators on the issue. (Before this time there were no indicators of potential political support or opposition to the WIPP project.)

The legislature killed the proposed referendum and once again we have a clear picture of the campaign for the next year and a half. We have generally reached our goals for this tactic. Avenues for releasing information have been established through newly created governor's, legislators', state and federal administrators' committees. Most importantly, Schlesinger has acceded to the states the right to veto waste disposal. While DOE probably thinks this eliminates an issue for their opposition; in fact, it gives the people — throughout the country — more sense of political efficacy, which will only increase their involvement.

6. Scientific

Our technical case against WIPP continues to gain strength as we explore the issues. Dr. Charles Hyder is circulating the draft of two papers on the geologic phenomenon called diapirism. Diapirism is upward-thrusting motion of ductile geologic material — salt, shale, clay, ice, ocean bottom sediments. These two papers outline one of our main technical objections to WIPP and are major generic criticisms of DOE's only serious approach to permanent waste disposal. Dr. Hyder will revise these and submit them for refereed publication in a month or so. In the meantime the papers have been sent to some 30 scientists around the country for comment.

The Sun Desert nuclear facility in California may be stopped due to a lack of adequate long term waste disposal. In January Dr. Hyder visited Gene Veranini and his technical staff of the California Energy Commission to present his findings. They are currently studying his papers in order to incorporate his conclusions in their final report. Dr. Hyder also contacted NRDC and Congressman McKloskey's office.

In addition to the arguments presented in Dr. Hyder's papers, we find the waste disposal plan vulnerable on the transportation/container safety issue. We have already obtained some relevant DOE information on this via the Freedom of Information Act and have written up some transportation research findings. Transportation/container safety are extremely salient to the people and we plan more related work on this topic in the near future.

We also have some pending research projects related to boom-town conditions in Carlsbad should the site be put in, and state financial burden if roads are upgraded to handle the waste traffic, as well as potential safety problems. These projects await funding for more research staff.

Hearings are now scheduled by the Nuclear Regulatory Commission (NRC), the state Environmental Improvement Agency, a state legislature standing committee, the Governor's Committee on Technical Excellence, the DOE, the Bureau of Land Management, and at least one congressional committee. We will present our technical data at every feasible opportunity.

We find that the technical information is necessary but not sufficient to stop

FIGURE 6 (continued)

WIPP. While we have good technical reasons for the infeasibility and hazards
of the project, we must be able to communicate this to New Mexicans, to convince them of the real danger, and organize them to oppose the project. Without
this effort, we feel that is very likely that the WIPP project will be implemented. As the industry understands — the true barriers to "permanent" waste
disposal are political and not technical. While our technical basis is solid
— much more work must be done in our media/research/organizing efforts. We
have limited funds to hire staff for the start of the campaign; we must obtain
more funds to continue salary obligations, to hire new legal and research
staff, and to produce and disseminate our media messages.

7. Organizing

Our organizing efforts are still small but are going well. We have two speakers
and one organizer working in southern New Mexico. In addition, we are in con-
tact with the anti-WIPP groups appearing throughout the state. Dr. Charles
Hyder has spoken in Albuquerque several times, in Santa Fe, and in Roswell.
Dr. Peter Montague has appeared in Carlsbad three times, in Albuquerque,
Cimmaron, Las Cruces, and Santa Fe.

We are now developing further strategies to focus anti-WIPP actions both locally
and state-wide. These approaches will become the core of our organizing efforts.
We are interviewing potential community/electoral campaign organizers to expand
the effort as soon as possible. The time-line for the campaign calls for hiring
two additional organizers by April 1. At that time we will be ready to announce
a state-wide campaign in at least ten communities, while really developing ways
to work in every community.

The future financial limitations mentioned previously affect our organizing
efforts. Media can be used most effectively in conjunction with local organizing, and media gives a critical legitimacy and presence that is necessary to
undertake local organizing. We will be feeling the pinch of limited funding
in our organizing efforts in the next few months.

8. Social Research

As mentioned, Lou Colombo has joined the staff recently and is completing our
initial research work. Lou, who specializes in survey research, computer
programming and statistical analysis, is preparing the way for a group experi-
enced in group process/qualitative research. Lou has contacted various
gubernatorial, congressional, and other state-wide campaigns plus various
"interest" or cultural groups such as Chicanos in northern and southern New
Mexico, farmers, etc. to understand the important groups in the state on the
WIPP issue. This was done to determine which groups will be contacted in our
first stage research. This research will entail structured group discussions
to determine how New Mexicans feel the radwaste issue affects their lives,
what they think about the issue, and the best ways to reach them with our information. The qualitative researchers are going to interview a selected
(stratified) sample of New Mexicans including farmers, miners, business people,
working class Chicanos, middle-class people, federal government workers, and
so on. Data from this research will be used in our organizing, media and
survey research efforts.

We have also investigated computer data sources and existing research related
to the waste disposal issue. In addition, different computerized data analysis
routines which will be used in the campaign have been located and will be implemented in the next few months.

9. Media

To date we have made some important impact on the media. As a result of a
Freedom of Information Act request we obtained information related to the WIPP
conceptual design studies. This information indicated that WIPP was designed

FIGURE 6 (continued)

to accept <u>all commercial and military wastes into the 21st century</u>. We publicized this to contrast with the DOE's claim that the site would contain only low and intermediate level military wastes. (They have applied for NRC licensing for high level military waste without "having made the decision" public that they plan to dispose of this waste in New Mexico.) Hence when DOE announces the intent to dispose of high level and commercial wastes, the campaign will be further legitimated.

In addition, we have publicized our attempt to obtain the preliminary Environmental Impact Statement under the Freedom of Information Act, and DOE's subsequent refusal to disclose it; and our technical argument that salt is dangerously mobile when high level wastes are placed in it. New Mexico newspapers, radio and TV have carried these statements. We have prepared a fact sheet to distribute in local organizing work, assisted in <u>La Lucha Nuclear</u> newsletter, and are just beginning a WIPP bi-monthly newsletter. Jeff is now writing a media handbook for local community organizers. As mentioned, we have made contact with a media production group. Right now we are just gearing up in terms of media. Jeff's arrival in mid-March will greatly facilitate our work here. The presence or absence of additional funding will make a great difference in our use of this very important information channel.

10. Legal

There are many ways in which WIPP is vulnerable to challenge and intervention such as on Nuclear Regulatory Commission or Department of Energy waste site criteria, Interstate Commerce Commission transportation criteria, the Environmental Impact Statement before the Department of Interior, the Environmental Impact Statement on the recently requested $40 million federal budget item for the initial development of WIPP, state legal actions, etc.

Our legal challenges to WIPP have, so far, been limited to freedom of information actions. This has included requests for internal documents and for the suppressed version of the Environmental Impact Statement. We have made personal contacts with both the Palo Alto and Washington, D.C. offices of the Natural Resources Defense Council and with Public Citizen. We are planning a legal strategy with these groups and others. We must have additional financial assistance to meet the costs of needed local legal work.

11. Conclusion

We hope this report provides a real impression of the current status of the anti-waste campaign in New Mexico. Much organizing work has already been done. The political environment is conducive to the campaign and promising for a nationally important win. However, we are really just beginning. We are laying a strong foundation for the campaign. In the next two months when the anti-waste movement does bloom, we will need additional funds to sustain the movement. A detailed description of our funding picture follows.

Source: Southwest Research and Information Center, P.O. Box 4524, Albuquerque, New Mexico 87106.

A number of recourses remain available to the concerned citizen even after the final agency decision has been made. The decision may be challenged in court, and pressure may still be brought to bear on the agency to reverse itself. At this late stage, however, much more effort will be required than was necessary before the agency reached its final decision. The citizen will want, in any case, to reassess why he lost and to carefully examine what alternatives are available at this juncture.

With some projects, a final decision at the conclusion of the impact statement process is effectively only the decision to proceed to the next stage of planning the project. For example, for highways, the environmental impact statement process is carried out at the corridor-approval stage. After the final environmental impact statement is written and a corridor selected, the regional administrator of the Federal Highway Administration must decide on the specific design of the highway. While a citizen may prefer that the highway project be dropped, a decision to go ahead with it, at this stage, means only a decision to proceed to further detailed planning. Accordingly, there still may be additional decision points "down the road," at which the citizen can object and attempt to defeat or modify the project, before actual construction and implementation.

A Final Word

The National Environmental Policy Act and the design of the environmental impact statement process encourage an unprecedented level of citizen involvement in government decision making. But what the ultimate effect of the citizens' use of the process will be is not yet clear.

This book has described the statute, outlined some possibilities, and reported some of what has happened so far. In the final analysis, however, the success of citizens in shaping a specific project will depend largely on the degree of interest and energy which they invest in their effort and the skill with which they make their views known. The impact statement process creates an opportunity for citizens to be heard and to influence a government decision. It does not guarantee an environmentally sound outcome.

The National Environmental Policy Act

THE NATIONAL ENVIRONMENTAL
POLICY ACT OF 1969, AS AMENDED*

An Act to establish a national policy for the environment, to provide for the establishment of a Council on Environmental Quality, and for other purposes.

Be it enacted by the Senate and House of Representatives of the United States of America in Congress assembled, That this Act may be cited as the "National Environmental Policy Act of 1969."

PURPOSE

SEC. 2. The purposes of this Act are: To declare a national policy which will encourage productive and enjoyable harmony between man and his environment; to promote efforts which will prevent or eliminate damage to the environment and biosphere and stimulate the health and welfare of man; to enrich the understanding of the ecological systems and natural resources important to the Nation; and to establish a Council on Environmental Quality.

TITLE I

DECLARATION OF NATIONAL ENVIRONMENTAL POLICY

SEC. 101. (a) The Congress, recognizing the profound impact of man's activity on the interrelations of all components of the natural environment, particularly the profound influences of population growth, high-density urbanization, industrial expansion, resource exploitation, and new and expanding technological advances and recognizing further the critical importance of restoring and maintaining environmental quality to the overall welfare and development of man, declares that it is the continuing policy of the Federal Government, in cooperation with State and local governments, and other concerned public and private organizations, to use all practicable means and measures, including financial and technical assistance, in a manner calculated to foster and promote the general welfare, to create and maintain conditions under which man and nature can exist in productive harmony, and fulfill the social, economic, and other requirements of present and future generations of Americans.

(b) In order to carry out the policy set forth in this Act, it is the continuing responsibility of the Federal Government to use all practicable means, consistent with other essential considerations of national policy, to improve

---

*Pub. L. 91–190, 42 U.S.C. 4321–4347, January 1, 1970, as amended by Pub. L. 94–83, August 9, 1975.

and coordinate Federal plans, functions, programs, and resources to the end that the Nation may—

(1) Fulfill the responsibilities of each generation as trustee of the environment for succeeding generations:

(2) Assure for all Americans safe, healthful, productive, and esthetically and culturally pleasing surroundings;

(3) Attain the widest range of beneficial uses of the environment without degradation, risk to health or safety, or other undesirable and unintended consequences;

(4) Preserve important historic, cultural, and natural aspects of our national heritage, and maintain, wherever possible, an environment which supports diversity, and variety of individual choice;

(5) Achieve a balance between population and resource use which will permit high standards of living and a wide sharing of life's amenities; and

(6) Enhance the quality of renewable resources and approach the maximum attainable recycling of depletable resources.

(c) The Congress recognizes that each person should enjoy a healthful environment and that each person has a responsibility to contribute to the preservation and enhancement of the environment.

Sec. 102. The Congress authorizes and directs that, to the fullest extent possible: (1) the policies, regulations, and public laws of the United States shall be interpreted and administered in accordance with the policies set forth in this Act, and (2) all agencies of the Federal Government shall—

(A) Utilize a systematic, interdisciplinary approach which will insure the integrated use of the natural and social sciences and the environmental design arts in planning and in decisionmaking which may have an impact on man's environment;

(B) Identify and develop methods and procedures, in consultation with the Council on Environmental Quality established by title II of this Act, which will insure that presently unquantified environmental amenities and values may be given appropirate consideration in decisionmaking along with economic and technical considerations;

(C) Include in every recommendation or report on proposals for legislation and other major Federal actions significantly affecting the quality of the human environment, a detailed statement by the responsible official on—

(i) The environmental impact of the proposed action,

(ii) Any adverse environmental effects which cannot be avoided should the proposal be implemented,

(iii) Alternatives to the proposed action,

(iv) The relationship between local short-term uses of man's environment and the maintenance and enhancement of long-term productivity, and

(v) Any irreversible and irretrievable commitments of resources which would be involved in the proposed action should it be implemented.

Prior to making any detailed statement, the responsible Federal official shall consult with and obtain the comments of any Federal agency which has jurisdiction by law or special expertise with respect to any environmental impact involved. Copies of such statement and the comments and views of the appropriate Federal, State, and local agencies, which are authorized to develop and enforce environmental standards, shall be made available to the President, the Council on Environmental Quality and to the public as provided by section 552 of title 5, United States Code, and shall accompany the proposal through the existing agency review processes;

(D) Any detailed statement required under subparagraph (C) after January 1, 1970, for any major Federal action funded under a program of grants to States shall not be deemed to be legally insufficient solely by reason of having been prepared by a State agency or official, if:

(i) the State agency or official has statewide jurisdiction and has the responsibility for such action,

(ii) the responsible Federal official furnishes guidance and participates in such preparation,

(iii) the responsible Federal official independently evaluates such statement prior to its approval and adoption, and

(iv) after January 1, 1976, the responsible Federal official provides early notification to, and solicits the views of, any other State or any Federal land management entity of any action or any alternative thereto which may have significant impacts upon such State or affected Federal land management entity and, if there is any disagreement on such impacts, prepares a written assessment of such impacts and views for incorporation into such detailed statement.

The procedures in this subparagraph shall not relieve the Federal official of his responsibilities for the scope, objectivity, and content of the entire statement or of any other responsibility under this Act; and further, this subparagraph does not affect the legal sufficiency of statements prepared by State agencies with less than statewide jurisdiction.

(E) Study, develop, and describe appropriate alternatives to recommended courses of action in any proposal which involves unresolved conflicts concerning alternative uses of available resources;

(F) Recognize the worldwide and long-range character of environmental problems and, where consistent with the foreign policy of the United States, lend appropriate support to initiatives, resolutions, and programs designed to maximize international cooperation in anticipating and preventing a decline in the quality of mankind's world environment;

(G) Make available to States, counties, municipalities, institutions, and individuals, advice and information useful in restoring, maintaining, and enhancing the quality of the environment;

(H) Initiate and utilize ecological information in the planning and development of resource-oriented projects; and

(I) Assist the Council on Environmental Quality established by title II of this Act.

SEC. 103. All agencies of the Federal Government shall review their present statutory authority, administrative regulations, and current policies and procedures for the purpose of determining whether there are any deficiencies or inconsistencies therein which prohibit full compliance with the purposes and provisions of this Act and shall propose to the President not later than July 1, 1971, such measures as may be necessary to bring their authority and policies into conformity with the intent, purposes, and procedures set forth in this Act.

SEC. 104. Nothing in section 102 or 103 shall in any way affect the specific statutory obligations of any Federal agency (1) to comply with criteria or standards of environmental quality, (2) to coordinate or consult with any other Federal or State agency, or (3) to act, or refrain from acting contingent upon the recommendations or certification of any other Federal or State agency.

SEC. 105. The policies and goals set forth in this Act are supplementary to those set forth in existing authorizations of Federal agencies.

TITLE II

COUNCIL ON ENVIRONMENTAL QUALITY

SEC. 201. The President shall transmit to the Congress annually beginning July 1, 1970, an Environmental Quality Report (hereinafter referred to as the "report") which shall set forth (1) the status and condition of the major natural, manmade, or altered environmental classes of the Nation, including, but not limited to, the air, the aquatic, including marine, estuarine, and fresh water, and the terrestrial environment, including, but not limited to, the forest, dryland, wetland, range, urban, suburban and rural environment; (2) current and foreseeable trends in the quality, management and utilization of such environments and the effects of those trends on the social, economic, and other requirements of the Nation; (3) the adequacy of available natural resources for fulfilling human and economic requirements of the Nation in the light of expected population pressures; (4) a review of the programs and activities (including regulatory activities) of the Federal Government, the State and local governments, and nongovernmental entities or individuals with particular reference to their effect on the environment and on the conservation, development and utilization of natural resources; and (5) a program for remedying the deficiencies of existing programs and activities, together with recommendations for legislation.

SEC. 202. There is created in the Executive Office of the President a Council on Environmental Quality (hereinafter referred to as the "Council"). The Council shall be composed of three members who shall be appointed by the President to serve at his pleasure, by and with the advice and consent of the Senate. The President shall designate one of the members of the Council to serve as Chairman. Each member shall be a person who, as a result of his training, experience, and attainments, is exceptionally well qualified to analyze and interpret environmental trends and information of all kinds; to appraise programs and activities of the Federal Government in the light of the policy set forth in title I of this Act; to be conscious of and responsive to the scientific, economic, social, esthetic, and cultural needs and interests of the Nation; and to formulate and recommend national policies to promote the improvement of the quality of the environment.

SEC. 203. The Council may employ such officers and employees as may be necessary to carry out its functions under this Act. In addition, the Council may employ and fix the compensation of such experts and consultants as may be necessary for the carrying out of its functions under this Act, in accordance with section 3109 of title 5, United States Code (but without regard to the last sentence thereof).

SEC. 204. It shall be the duty and function of the Council—

(1) To assist and advise the President in the preparation of the Environmental Quality Report required by section 201;

(2) To gather timely and authoritative information concerning the conditions and trends in the quality of the environment both current and prospective, to analyze and interpret such information for the purpose of determining whether such conditions and trends are interfering, or are likely to interfere, with the achievement of the policy set forth in title I of this Act, and to compile and submit to the President studies relating to such conditions and trends;

(3) To review and appraise the various programs and activities of the Federal Government in the light of the policy set forth in title I of this Act for the purpose of determining the extent to which such programs and activities are contributing to the achievement of such policy, and to make recommendations to the President with respect thereto;

(4) To develop and recommend to the President national policies to foster and promote the improvement of environmental quality to meet the conservation, social, economic, health, and other requirements and goals of the Nation;

(5) To conduct investigations, studies, surveys, research, and analyses relating to ecological systems and environmental quality;

(6) To document and define changes in the natural environment, including the plant and animal systems, and to accumulate necessary

data and other information for a continuing analysis of these changes or trends and an interpretation of their underlying causes;

(7) To report at least once each year to the President on the state and condition of the environment; and

(8) To make and furnish such studies, reports thereon, and recommendations with respect to matters of policy and legislation as the President may request.

SEC. 205. In exercising its powers, functions, and duties under this Act, the Council shall—

(1) Consult with the Citizens' Advisory Committee on Environmental Quality established by Executive Order No. 11472, dated May 29, 1969, and with such representatives of science, industry, agriculture, labor, conservation organizations, State and local governments and other groups, as it deems advisable; and

(2) Utilize, to the fullest extent possible, the services, facilities and information (including statistical information) of public and private agencies and organizations, and individuals, in order that duplication of effort and expense may be avoided, thus assuring that the Council's activities will not unnecessarily overlap or conflict with similar activities authorized by law and performed by established agencies.

SEC. 206. Members of the Council shall serve full time and the Chairman of the Council shall be compensated at the rate provided for Level II of the Executive Schedule Pay Rates (5 U.S.C. 5313). The other members of the Council shall be compensated at the rate provided for Level IV of the Executive Schedule Pay Rates (5 U.S.C. 5315).

SEC. 207. There are authorized to be appropriated to carry out the provisions of this Act not to exceed $300,000 for fiscal year 1970, $700,000 for fiscal year 1971, and $1 million for each fiscal year thereafter.

Approved January 1, 1970.

B

CEQ Guidelines on Preparation of Environmental Impact Statements

PREPARATION OF ENVIRONMENTAL IMPACT STATEMENTS: GUIDELINES*

On May 2, 1973, the Council on Environmental Quality published in the FEDERAL REGISTER, for public comment, a proposed revision of its guidelines for the preparation of environmental impact statements. Pursuant to the National Environmental Policy Act (P.L. 91–190, 42 U.S.C. 4321 et seq.) and Executive Order 11514 (35 FR 4247) all Federal departments, agencies, and establishments are required to prepare such statements in connection with their proposals for legislation and other major Federal actions significantly affecting the quality of the human environment. The authority for the Council's guidelines is set forth below in § 1500.1. The specific policy to be implemented by the guidelines is set forth below in § 1500.2.

The Council received numerous comments on its proposed guidelines from environmental groups, Federal, State, and local agencies, industry, and private individuals. Two general themes were presented in the majority of the comments. First, the Council should increase the opportunity for public involvement in the impact statement process. Second, the Council should provide more detailed guidance on the responsibilities of Federal agencies in light of recent court decisions interpreting the Act. The proposed guidelines have been revised in light of the specific comments relating to these general themes, as well as other comments received, and are now being issued in final form.

The guidelines will appear in the Code of Federal Regulations in Title 40, Chapter V, at Part 1500. They are being codified, in part, because they affect State and local governmental agencies, environmental groups, industry, and private individuals, in addition to Federal agencies, to which they are specifically directed, and the resultant need to make them widely and readily available.

Sec.
1500.1   Purpose and authority.
1500.2   Policy.
1500.3   Agency and OMB procedures.
1500.4   Federal agencies included: effect of the act on existing agency mandates.
1500.5   Types of actions covered by the act.
1500.6   Identifying major actions significantly affecting the environment.
1500.7   Preparing draft environmental statements; public hearings.
1500.8   Content of environmental statements.
1500.9   Review of draft environmental statements by Federal, Federal-State, State, and local agencies and by the public.
1500.10   Preparation and circulation of final environmental statements.
1500.11   Transmittal of statements to the Council; minimum periods for review; requests by the Council.

---

*38 Fed. Reg. 20550–20562 (1973).

AUTHORITY: National Environmental Policy Act (P.L. 91–190, 42 U.S.C.
4321 et seq.) and Executive Order 11514.

§ 1500.1   PURPOSE AND AUTHORITY

(a) This directive provides guidelines to Federal departments, agencies, and
establishments for preparing detailed environmental statements on proposals
for legislation and other major Federal actions significantly affecting the
quality of the human environment as required by section 102(2)(C) of the
National Environmental Policy Act (P.L. 91–190, 42 U.S.C. 4321 et seq.,
hereafter "the Act"). Underlying the preparation of such environmental
statements is the mandate of both the Act and Executive Order 11514 (35 FR
4247) of March 5, 1970, that all Federal agencies, to the fullest extent possible, direct their policies, plans and programs to protect and enhance environmental quality. Agencies are required to view their actions in a manner
calculated to encourage productive and enjoyable harmony between man
and his environment, to promote efforts preventing or eliminating damage to
the environment and biosphere and stimulating the health and welfare of
man, and to enrich the understanding of the ecological systems and natural
resources important to the Nation. The objective of section 102(2)(C) of
the Act and of these guidelines is to assist agencies in implementing these
policies. This requires agencies to build into their decisionmaking process, beginning at the earliest possible point, an appropriate and careful consideration of the environmental aspects of proposed action in order that adverse
environmental effects may be avoided or minimized and environmental quality previously lost may be restored. This directive also provides guidance to
Federal, State, and local agencies and the public in commenting on statements prepared under these guidelines.

(b) Pursuant to section 204(3) of the Act the Council on Environmental
Quality (hereafter "the Council") is assigned the duty and function of reviewing and appraising the programs and activities of the Federal Government,
in the light of the Act's policy, for the purpose of determining the extent to
which such programs and activities are contributing to the achievement of
such policy, and to make recommendations to the President with respect
thereto. Section 102(2)(B) of the Act directs all Federal agencies to identify
and develop methods and procedures, in consultation with the Council, to
insure that unquantified environmental values be given appropriate consideration in decisionmaking along with economic and technical considerations;
section 102(2)(C) of the Act directs that copies of all environmental impact
statements be filed with the Council; and section 102(2)(H) directs all
Federal agencies to assist the Council in the performance of its functions.
These provisions have been supplemented in sections 3 (h) and (i) of Executive Order 11514 by direction that the Council issue guidelines to Federal

agencies for preparation of environmental impact statements and such other instructions to agencies and requests for reports and information as may be required to carry out the Council's responsibilities under the Act.

§ 1500.2   POLICY

(a) As early as possible and in all cases prior to agency decision concerning recommendations or favorable reports on proposals for (1) legislation significantly affecting the quality of the human environment (see §§ 1500.5(i) and 1500.12) (hereafter "legislative actions") and (2) all other major Federal actions significantly affecting the quality of the human environment (hereafter "administrative actions"), Federal agencies will, in consultation with other appropriate Federal, State and local agencies and the public assess in detail the potential environmental impact.

(b) Initial assessments of the environmental impacts of proposed action should be undertaken concurrently with initial technical and economic studies and, where required, a draft environmental impact statement prepared and circulated for comment in time to accompany the proposal through the existing agency review processes for such action. In this process, Federal agencies shall: (1) Provide for circulation of draft environmental statements to other Federal, State, and local agencies and for their availability to the public in accordance with the provisions of these guidelines; (2) consider the comments of the agencies and the public; and (3) issue final environmental impact statements responsive to the comments received. The purpose of this assessment and consultation process is to provide agencies and other decision-makers as well as members of the public with an understanding of the potential environmental effects of proposed actions, to avoid or minimize adverse effects wherever possible, and to restore or enhance environmental quality to the fullest extent practicable. In particular, agencies should use the environmental impact statement process to explore alternative actions that will avoid or minimize adverse impacts and to evaluate both the long- and short-range implications of proposed actions to man, his physical and social surroundings, and to nature. Agencies should consider the results of their environmental assessments along with their assessments of the net economic, technical and other benefits of proposed actions and use all practicable means, consistent with other essential considerations of national policy, to restore environmental quality as well as to avoid or minimize undesirable consequences for the environment.

§ 1500.3   AGENCY AND OMB PROCEDURES

(a) Pursuant to section 2(f) of Executive Order 11514, the heads of Federal agencies have been directed to proceed with measures required by section 102 (2)(C) of the Act. Previous guidelines of the Council directed each agency to establish its own formal procedures for (1) identifying those agency actions requiring environmental statements, the appropriate time prior to decision for the consultations required by section 102(2)(C) and the agency review process for which environmental statements are to be available, (2) obtaining information required in their preparation, (3) designating the officials who are to be responsible for the statements, (4) consulting with and taking account of the comments of appropriate Federal, State and local agencies and the public, including obtaining the comment of the Administrator of the Environmental Protection Agency when required under section 309 of the Clean Air Act, as amended, and (5) meeting the requirements of section 2(b) of Executive Order 11514 for providing timely public information on Federal plans and programs with environmental impact. Each agency, including both departmental and subdepartmental components having such procedures, shall review its procedures and shall revise them, in consultation with the Council, as may be necessary in order to respond to requirements imposed by these revised guidelines as well as by such previous directions. After such consultation, proposed revisions of such agency procedures shall be published in the FEDERAL REGISTER no later than October 30, 1973. A minimum 45-day period for public comment shall be provided, followed by publication of final procedures no later than forty-five (45 days) after the conclusion of the comment period. Each agency shall submit seven (7) copies of all such procedures to the Council. Any future revision of such agency procedures shall similarly be proposed and adopted only after prior consultation with the Council and, in the case of substantial revision, opportunity for public comment. All revisions shall be published in the FEDERAL REGISTER.

(b) Each Federal agency should consult, with the assistance of the Council and the Office of Management and Budget if desired, with other appropriate Federal agencies in the development and revision of the above procedures so as to achieve consistency in dealing with similar activities and to assure effective coordination among agencies in their review of proposed activities. Where applicable State and local review of such agency procedures should be conducted pursuant to procedures established by Office of Management and Budget Circular No. A–85.

(c) Existing mechanisms for obtaining the views of Federal, State, and local agencies on proposed Federal actions should be utilized to the maximum extent practicable in dealing with environmental matters. The Office of Management and Budget will issue instructions, as necessary, to take full advantage of such existing mechanisms.

§ 1500.4 FEDERAL AGENCIES INCLUDED; EFFECT OF THE ACT ON EXISTING AGENCY MANDATES

(a) Section 102(2)(C) of the Act applies to all agencies of the Federal Government. Section 102 of the Act provides that "to the fullest extent possible: (1) The policies, regulations, and public laws of the United States shall be interpreted and administered in accordance with the policies set forth in this Act," and section 105 of the Act provides that "the policies and goals set forth in this Act are supplementary to those set forth in existing authorizations of Federal agencies." This means that each agency shall interpret the provisions of the Act as a supplement to its existing authority and as a mandate to view traditional policies and missions in the light of the Act's national environmental objectives. In accordance with this purpose, agencies should continue to review their policies, procedures, and regulations and to revise them as necessary to ensure full compliance with the purposes and provisions of the Act. The phrase "to the fullest extent possible" in section 102 is meant to make clear that each agency of the Federal Government shall comply with that section unless existing law applicable to the agency's operations expressly prohibits or makes compliance impossible.

§ 1500.5 TYPES OF ACTIONS COVERED BY THE ACT

(a) "Actions" include but are not limited to:

(1) Recommendations or favorable reports relating to legislation including requests for appropriations. The requirement for following the section 102 (2)(C) procedure as elaborated in these guidelines applies to both (i) agency recommendations on their own proposals for legislation (see § 1500.12); and (ii) agency reports on legislation initiated elsewhere. In the latter case only the agency which has primary responsibility for the subject matter involved will prepare an environmental statement.

(2) New and continuing projects and program activities: directly undertaken by Federal agencies; or supported in whole or in part through Federal contracts, grants, subsidies, loans, or other forms of funding assistance (except where such assistance is solely in the form of general revenue sharing funds, distributed under the State and Local Fiscal Assistance Act of 1972, 31 U.S.C. 1221 et seq. with no Federal agency control over the subsequent use of such funds); or involving a Federal lease, permit, license certificate or other entitlement for use.

(3) The making, modification, or establishment of regulations, rules, procedures, and policy.

§ 1500.6　　IDENTIFYING MAJOR ACTIONS SIGNIFICANTLY AFFECTING THE ENVIRONMENT

(a) The statutory clause "major Federal actions significantly affecting the quality of the human environment" is to be construed by agencies with a view to the overall, cumulative impact of the action proposed, related Federal actions and projects in the area, and further actions contemplated. Such actions may be localized in their impact, but if there is potential that the environment may be significantly affected, the statement is to be prepared. Proposed major actions, the environmental impact of which is likely to be highly controversial, should be covered in all cases. In considering what constitutes major action significantly affecting the environment, agencies should bear in mind that the effect of many Federal decisions about a project or complex of projects can be individually limited but cumulatively considerable. This can occur when one or more agencies over a period of years puts into a project individually minor but collectively major resources, when one decision involving a limited amount of money is a precedent for action in much larger cases or represents a decision in principle about a future major course of action, or when several Government agencies individually make decisions about partial aspects of a major action. In all such cases, an environmental statement should be prepared if it is reasonable to anticipate a cumulatively significant impact on the environment from Federal action. The Council, on the basis of a written assessment of the impacts involved, is available to assist agencies in determining whether specific actions require impact statements.

(b) Section 101(b) of the Act indicates the broad range of aspects of the environment to be surveyed in any assessment of significant effect. The Act also indicates that adverse significant effects include those that degrade the quality of the environment, curtail the range of beneficial uses of the environment, and serve short-term, to the disadvantage of long-term, environmental goals. Significant effects can also include actions which may have both beneficial and detrimental effects, even if on balance the agency believes that the effect will be beneficial. Significant effects also include secondary effects, as described more fully, for example, in § 1500.8(a)(iii)(B). The significance of a proposed action may also vary with the setting, with the result that an action that would have little impact in an urban area may be significant in a rural setting or vice versa. While a precise definition of environmental "significance," valid in all contexts, is not possible, effects to be considered in assessing significance include, but are not limited to, those outlined in Appendix II of these guidelines.

(c) Each of the provisions of the Act, except section 102(2)(C), applies to all Federal agency actions. Section 102(2)(C) requires the preparation of a detailed environmental impact statement in the case of "major Federal actions significantly affecting the quality of the human environment." The identification of major actions significantly affecting the environment is the responsibility of each Federal agency, to be carried out against the background of its own particular operations. The action must be a (1) "major" action, (2) which is a "Federal action," (3) which has a "significant" effect, and (4) which involves the "quality of the human environment." The words "major" and "significantly" are intended to imply thresholds of importance and impact that must be met before a statement is required. The action causing the impact must also be one where there is sufficient Federal control and responsibility to constitute "Federal action" in contrast to cases where such Federal control and responsibility are not present as, for example, when Federal funds are distributed in the form of general revenue sharing to be used by State and local governments (see § 1500.5a(ii)). Finally, the action must be one that significantly affects the quality of the human environment either by directly affecting human beings or by indirectly affecting human beings through adverse effects on the environment. Each agency should review the typical classes of actions that it undertakes and, in consultation with the Council, should develop specific criteria and methods for identifying those actions likely to require environmental statements and those actions likely not to require environmental statements. Normally this will involve:

(i) Making an initial assessment of the environmental impacts typically associated with principal types of agency action.

(ii) Identifying on the basis of this assessment, types of actions which normally do, and types of actions which normally do not, require statements.

(iii) With respect to remaining actions that may require statements depending on the circumstances, and those actions determined under the preceding paragraph (ii) of this section as likely to require statements, identifying: (*a*) what basic information needs to be gathered; (*b*) how and when such information is to be assembled and analyzed; and (*c*) on what bases environmental assessments and decisions to prepare impact statements will be made. Agencies may either include this substantive guidance in the procedures issued pursuant to § 1500.3(a) of these guidelines, or issue such guidance as supplemental instructions to aid relevant agency personnel in implementing the impact statement process. Pursuant to § 1500.14 of these guidelines, agencies shall report to the Council by June 30, 1974, on the progress made in developing such substantive guidance.

(d)(1) Agencies should give careful attention to identifying and defining the purpose and scope of the action which would most appropriately serve as the subject of the statement. In many cases, broad program statements will be required in order to assess the environmental effects of a number of individual actions on a given geographical area (e.g., coal leases), or environmental impacts that are generic or common to a series of agency actions (e.g., maintenance or waste handling practices), or the overall impact of a large-scale program or chain of contemplated projects (e.g., major lengths of highway as opposed to small segments). Subsequent statements on major individual actions will be necessary where such actions have significant environmental impacts not adequately evaluated in the program statement.

(2) Agencies engaging in major technology research and development programs should develop procedures for periodic evaluation to determine when a program statement is required for such programs. Factors to be considered in making this determination include the magnitude of Federal investment in the program, the likelihood of widespread application of the technology, the degree of environmental impact which would occur if the technology were widely applied, and the extent to which continued investment in the new technology is likely to restrict future alternatives. Statements must be written late enough in the development process to contain meaningful information, but early enough so that this information can practically serve as an input in the decision-making process. Where it is anticipated that a statement may ultimately be required but that its preparation is still premature, the agency should prepare an evaluation briefly setting forth the reasons for its determi-

nation that a statement is not yet necessary. This evaluation should be periodically updated, particularly when significant new information becomes available concerning the potential environmental impact of the program. In any case, a statement must be prepared before research activities have reached a stage of investment or commitment to implementation likely to determine subsequent development or restrict later alternatives. Statements on technology research and development programs should include an analysis not only of alternative forms of the same technology that might reduce any adverse environmental impacts but also of alternative technologies that would serve the same function as the technology under consideration. Efforts should be made to involve other Federal agencies and interested groups with relevant expertise in the preparation of such statements because the impacts and alternatives to be considered are likely to be less well defined than in other types of statements.

(e) In accordance with the policy of the Act and Executive Order 11514 agencies have a responsibility to develop procedures to insure the fullest practicable provision of timely public information and understanding of Federal plans and programs with environmental impact in order to obtain the views of interested parties. In furtherance of this policy, agency procedures should include an appropriate early notice system for informing the public of the decision to prepare a draft environmental statement on proposed administrative actions (and for soliciting comments that may be helpful in preparing the statement) as soon as is practicable after the decision to prepare the statement is made. In this connection, agencies should: (1) maintain a list of administrative actions for which environmental statements are being prepared; (2) revise the list at regular intervals specified in the agency's procedures developed pursuant to § 1500.3(a) of these guidelines (but not less than quarterly) and transmit each such revision to the Council; and (3) make the list available for public inspection on request. The Council will periodically publish such lists in the FEDERAL REGISTER. If an agency decides that an environmental statement is not necessary for a proposed action (i) which the agency has identified pursuant to § 1500.6(c)(ii) as normally requiring preparation of a statement, (ii) which is similar to actions for which the agency has prepared a significant number of statements, (iii) which the agency has previously announced would be the subject of a statement, or (iv) for which the agency has made a negative determination in response to a request from the Council pursuant to § 1500.11(f), the agency shall prepare a publicly available record briefly setting forth the agency's decision and the reasons for that determination. Lists of such negative determinations, and any evaluations made pursuant to § 1500.6 which conclude that preparation of a statement is not yet timely, shall be prepared and made available in the same manner as provided in this subsection for lists of statements under preparation.

§ 1500.7   PREPARING DRAFT ENVIRONMENTAL STATEMENTS; PUBLIC HEARINGS

(a) Each environmental impact statement shall be prepared and circulated in draft form for comment in accordance with the provisions of these guidelines. The draft statement must fulfill and satisfy to the fullest extent possible at the time the draft is prepared the requirements established for final statements by section 102(2)(C). (Where an agency has an established practice of declining to favor an alternative until public comments on a proposed action have been received, the draft environmental statement may indicate that two or more alternatives are under consideration.) Comments received shall be carefully evaluated and considered in the decision process. A final statement with substantive comments attached shall then be issued and circulated in accordance with applicable provisions of §§ 1500.10, 1500.11, or 1500.12. It is important that draft environmental statements be prepared and circulated for comment and furnished to the Council as early as possible in the agency review process in order to permit agency decisionmakers and outside reviewers to give meaningful consideration to the environmental issues involved. In particular, agencies should keep in mind that such statements are to serve as the means of assessing the environmental impact of proposed agency actions, rather than as a justification for decisions already made. This means that draft statements on administrative actions should be prepared and circulated for comment prior to the first significant point of decision in the agency review process. For major categories of agency action, this point should be identified in the procedures issued pursuant to § 1500.3(a). For major categories of projects involving an applicant and identified pursuant to § 1500.6(c)(ii) as normally requiring the preparation of a statement, agencies should include in their procedures provisions limiting actions which an applicant is permitted to take prior to completion and review of the final statement with respect to his application.

(b) Where more than one agency (1) directly sponsors an action, or is directly involved in an action through funding, licenses, or permits, or (2) is involved in a group of actions directly related to each other because of their functional interdependence and geographical proximity, consideration should be given to preparing one statement for all the Federal actions involved (see § 1500.6(d)(1)). Agencies in such cases should consider the possibility of joint preparation of a statement by all agencies concerned, or designation of a single "lead agency" to assume supervisory responsibility for preparation of the statement. Where a lead agency prepares the statement, the other agencies involved should provide assistance with respect to their areas of jurisdiction and expertise. In either case, the statement should contain an environmental assessment of the full range of Federal actions involved, should reflect the views of all participating agencies, and should be prepared before major or irreversible actions have been taken by any of the participating agencies. Factors relevant in determining an appropriate lead agency include the time sequence in which the agencies become involved, the magnitude of their respective involvement, and their relative expertise with respect to the project's environmental effects. As necessary, the Council will assist in resolving questions of responsibility for statement preparation in the case of multi-agency actions. Federal Regional Councils, agencies and the public are encouraged to bring to the attention of the Council and other relevant agencies appropriate situations where a geographic or regionally focused statement would be desirable because of the cumulative environmental effects likely to result from multi-agency actions in the area.

(c) Where an agency relies on an applicant to submit initial environmental information, the agency should assist the applicant by outlining the types of information required. In all cases, the agency should make its own evaluation of the environmental issues and take responsibility for the scope and content of draft and final environmental statements.

(d) Agency procedures developed pursuant to § 1500.3(a) of these guidelines should indicate as explicitly as possible those types of agency decisions or actions which utilize hearings as part of the normal agency review process, either as a result of statutory requirement or agency practice. To the fullest extent possible, all such hearings shall include consideration of the environmental aspects of the proposed action. Agency procedures shall also specifically include provision for public hearings on major actions with environmental impact, whenever appropriate, and for providing the public with relevant information, including information on alternative courses of action. In deciding whether a public hearing is appropriate, an agency should consider: (1) the magnitude of the proposal in terms of economic costs, the geographic area involved, and the uniqueness or size of commitment of the resources involved; (2) the degree of interest in the proposal, as evidenced by requests from the public and from Federal, State and local authorities that a hearing be held; (3) the complexity of the issue and the likelihood that information will be presented at the hearing which will be of assistance to the agency in fulfilling its responsibilities under the Act; and (4) the extent to which public involvement already has been achieved through other means, such as earlier public hearings, meetings with citizen representatives, and/or written comments on the proposed action. Agencies should make any draft environmental statements to be issued available to the public at least fifteen (15) days prior to the time of such hearings.

§ 1500.8   CONTENT OF ENVIRONMENTAL STATEMENTS

(a) The following points are to be covered:

(1) A description of the proposed action, a statement of its purposes, and a description of the environment affected, including information, summary technical data, and maps and diagrams where relevant, adequate to permit an assessment of potential environmental impact by commenting agencies and the public. Highly technical and specialized analyses and data should be avoided in the body of the draft impact statement. Such materials should be attached as appendices or footnoted with adequate bibliographic references. The statement should also succinctly describe the environment of the area affected as it exists prior to a proposed action, including other Federal activities in the area affected by the proposed action which are related to the proposed action. The interrelationships and cumulative environmental impacts of the proposed action and other related Federal projects shall be presented in the statement. The amount of detail provided in such descriptions should be commensurate with the extent and expected impact of the action, and with the amount of information required at the particular level of decisionmaking (planning, feasibility, design, etc.). In order to ensure accurate descriptions and environmental assessments, site visits should be made where feasible. Agencies should also take care to identify, as appropriate, population and growth characteristics of the affected area and any population and growth assumptions used to justify the project or program or to determine secondary population and growth impacts resulting from the proposed action and its alternatives (see paragraph (3)(ii), of this section). In discussing these population aspects, agencies should give consideration to using the rates of growth in the region of the project contained in the projection compiled for the Water Resources Council by the Bureau of Economic Analysis of the Department of Commerce and the Economic Research Service of the Department of Agriculture (the "OBERS" projection). In any event it is essential that the sources of data used to identify, quantify or evaluate any and all environmental consequences be expressly noted.

(2) The relationship of the proposed action to land use plans, policies, and controls for the affected area. This requires a discussion of how the proposed action may conform or conflict with the objectives and specific terms of approved or proposed Federal, State, and local land use plans, policies and controls, if any, for the area affected, including those developed in response to the Clean Air Act or the Federal Water Pollution Control Act Amendments of 1972. Where a conflict or inconsistency exists, the statement should describe the extent to which the agency has reconciled its proposed action with the plan, policy or control and the reasons why the agency has decided to proceed notwithstanding the absence of full reconciliation.

(3) The probable impact of the proposed action on the environment.

(i) This requires agencies to assess the positive and negative effects of the proposed action as it affects both the national and international environment. The attention given to different environmental factors will vary according to the nature, scale, and location of proposed actions. Among factors to consider should be the potential effect of the action on such aspects of the environment as those listed in Appendix II of these guidelines. Primary attention should be given in the statement to discussing those factors most evidently impacted by the proposed action.

(ii) Secondary or indirect, as well as primary or direct, consequences for the environment should be included in the analysis. Many major Federal actions, in particular those that involve the construction or licensing of infrastructure investments (e.g., highways, airports, sewer systems, water resource projects, etc.), stimulate or induce secondary effects in the form of associated investments and changed patterns of social and economic activities. Such secondary effects, through their impacts on existing community facilities and activities, through inducing new facilities and activities, or through changes in natural conditions, may often be even more substantial than the primary effects of the original action itself. For example, the effects of the proposed action on population and growth may be among the more significant secondary effects. Such population and growth impacts should be estimated if expected to be significant (using data identified as indicated in § 1500.8(a)(1)) and an assessment made of the effect of any possible change in population patterns or growth upon the resource base, including land use, water, and public services, of the area in question.

(4) Alternatives to the proposed action, including, where relevant, those not within the existing authority of the responsible agency. (Section 102(2)(D) of the Act requires the responsible agency to "study, develop, and describe appropriate alternatives to recommended courses of action in any proposal which involves unresolved conflicts concerning alternative uses of available resources"). A rigorous exploration and objective evaluation of the environmental impacts of all reasonable alternative actions, particularly those that might enhance environmental quality or avoid some or all of the adverse environmental effects, is essential. Sufficient analysis of such alternatives and their environmental benefits, costs and risks should accompany the proposed action through the agency review process in order not to foreclose prematurely options which might enhance environmental quality or have less detrimental effects. Examples of such alternatives include: the alternative of taking no action or of postponing action pending further study; alternatives requiring actions of a significantly different nature which would provide similar benefits with different environmental impacts (e.g., nonstructural alternatives to flood control programs, or mass transit alternatives to highway construction); alternatives related to different designs or details of the proposed action which would present different environmental impacts (e.g., cooling ponds vs. cooling towers for a power plant or alternatives that will significantly conserve energy); alternative measures to provide for compensation of fish and wildlife losses, including the acquisition of land, waters, and interests therein. In each case, the analysis should be sufficiently detailed to reveal the agency's comparative evaluation of the environmental benefits, costs and risks of the proposed action and each reasonable alternative. Where an existing impact statement already contains such an analysis, its treatment of alternatives may be incorporated provided that such treatment is current and relevant to the precise purpose of the proposed action.

(5) Any probable adverse environmental effects which cannot be avoided (such as water or air pollution, undesirable land use patterns, damage to life systems, urban congestion, threats to health or other consequences adverse to the environmental goals set out in section 101(b) of the Act). This should be a brief section summarizing in one place those effects discussed in paragraph (a)(3) of this section that are adverse and unavoidable under the proposed action. Included for purposes of contrast should be a clear statement of how other avoidable adverse effects discussed in paragraph (a)(2) of this section will be mitigated.

(6) The relationship between local short-term uses of man's environment and the maintenance and enhancement of long-term productivity. This section should contain a brief discussion of the extent to which the proposed action involves tradeoffs between short-term environmental gains at the expense

of long-term losses, or vice versa, and a discussion of the extent to which the proposed action forecloses future options. In this context short-term and long-term do not refer to any fixed time periods, but should be viewed in terms of the environmentally significant consequences of the proposed action.

(7) Any irreversible and irretrievable commitments of resources that would be involved in the proposed action should it be implemented. This requires the agency to identify from its survey unavoidable impacts in paragraph (a)(5) of this section the extent to which the action irreversibly curtails the range of potential uses of the environment. Agencies should avoid construing the term "resources" to mean only the labor and materials devoted to an action. "Resources" also means the natural and cultural resources committed to loss or destruction by the action.

(8) An indication of what other interests and considerations of Federal policy are thought to offset the adverse environmental effects of the proposed action identified pursuant to paragraphs (a)(3) and (5) of this section. The statement should also indicate the extent to which these stated countervailing benefits could be realized by following reasonable alternatives to the proposed action (as identified in paragraph (a)(4) of this section) that would avoid some or all of the adverse environmental effects. In this connection, agencies that prepare cost-benefit analyses of proposed actions should attach such analyses, or summaries thereof, to the environmental impact statement, and should clearly indicate the extent to which environmental costs have not been reflected in such analyses.

(b) In developing the above points agencies should make every effort to convey the required information succinctly in a form easily understood, both by members of the public and by public decisionmakers, giving attention to the substance of the information conveyed rather than to the particular form, or length, or detail of the statement. Each of the above points, for example, need not always occupy a distinct section of the statement if it is otherwise adequately covered in discussing the impact of the proposed action and its alternatives—which items should normally be the focus of the statement. Draft statements should indicate at appropriate points in the text any underlying studies, reports, and other information obtained and considered by the agency in preparing the statement including any cost-benefit analyses prepared by the agency, and reports of consulting agencies under the Fish and Wildlife Coordination Act, 16 U.S.C. 661 et seq., and the National Historic Preservation Act of 1966, 16 U.S.C. 470 et seq., where such consultation has taken place. In the case of documents not likely to be easily accessible (such as internal studies or reports), the agency should indicate how such information may be obtained. If such information is attached to the statement, care should be taken to ensure that the statement remains an essentially self-contained instrument, capable of being understood by the reader without the need for undue cross reference.

(c) Each environmental statement should be prepared in accordance with the precept in section 102(2)(A) of the Act that all agencies of the Federal Government "utilize a systematic, interdisciplinary approach which will insure the integrated use of the natural and social sciences and the environmental design arts in planning and decisionmaking which may have a impact on man's environment." Agencies should attempt to have relevant disciplines represented on their own staffs; where this is not feasible they should make appropriate use of relevant Federal, State, and local agencies or the professional services of universities and outside consultants. The interdisciplinary approach should not be limited to the preparation of the environmental impact statement, but should also be used in the early planning stages of the proposed action. Early application of such an approach should help assure a systematic evaluation of reasonable alternative courses of action and their potential social, economic, and environmental consequences.

(d) Appendix I prescribes the form of the summary sheet which should accompany each draft and final environmental statement.

§ 1500.9   REVIEW OF DRAFT ENVIRONMENTAL STATEMENTS BY FEDERAL, FEDERAL-STATE, STATE, AND LOCAL AGENCIES AND BY THE PUBLIC

(a) *Federal agency review: In general.* A Federal agency considering an action requiring an environmental statement should consult with, and (on the basis of a draft environmental statement for which the agency takes responsibility) obtain the comment on the environmental impact of the action of Federal and Federal-State agencies with jurisdiction by law or special expertise with respect to any environmental impact involved. These Federal and Federal-State agencies and their relevant areas of expertise include those identified in Appendices II and III to these guidelines. It is recommended that the listed departments and agencies establish contact points, which may be regional offices, for providing comments on the environmental statements. The requirement in section 102(2)(C) to obtain comment from Federal agencies having jurisdiction or special expertise is in addition to any specific statutory obligation of any Federal agency to coordinate or consult with any other Federal or State agency. Agencies should, for example, be alert to consultation requirements of the Fish and Wildlife Coordination Act, 16 U.S.C. 661 et seq., and the National Historic Preservation Act of 1966, 16 U.S.C. 470 et seq. To the extent possible, statements or findings concerning environmental impact required by other statutes, such as section 4(f) of the Department of Transportation Act of 1966, 49 U.S.C. 1653(f), or section 106 of the National Historic Preservation Act of 1966, should be combined with compliance with the environmental impact statement requirements of section 102(2)(C) of the Act to yield a single document which meets all applicable requirements. The Advisory Council on Historic Preservation, the Department of Transportation, and the Department of Interior, in consultation with the Council, will issue any necessary supplementing instructions for furnishing information or findings not forthcoming under the environmental impact statement process.

(b) *EPA review:* Section 309 of the Clean Air Act, as amended (42 U.S.C. § 1857h–7), provides that the Administrator of the Environmental Protection Agency shall comment in writing on the environmental impact of any matter relating to his duties and responsibilities, and shall refer to the Council any matter that the Administrator determines is unsatisfactory from the standpoint of public health or welfare or environmental quality. Accordingly, wherever an agency action related to air or water quality, noise abatement and control, pesticide regulation, solid waste disposal, generally applicable environmental radiation criteria and standards, or other provision of the authority of the Administrator involved, Federal agencies are required to submit such proposed actions and their environmental impact statements, if such have been prepared, to the Administrator for review and comment in writing. In all cases where EPA determines that proposed agency action is environmentally unsatisfactory, or where EPA determines that an environmental statement is so inadequate that such a determination cannot be made, EPA shall publish its determination and notify the Council as soon as practicable. The Administrator's comments shall constitute his comments for the purposes of both section 309 of the Clean Air Act and section 102(2)(C) of the National Environmental Policy Act.

(c) *State and local review:* Office of Management and Budget Circular No. A–95 (Revised) through its system of State and areawide clearinghouses provides a means for securing the views of State and local environmental agencies, which can assist in the preparation and review of environmental impact statements. Current instructions for obtaining the views of such agencies are contained in the joint OMB–CEQ memorandum attached to these guidelines as Appendix IV. A current listing of clearinghouses is issued periodically by the Office of Management and Budget.

(d) *Public review:* The procedures established by these guidelines are designed to encourage public participation in the impact statement process at the earliest possible time. Agency procedures should make provision for facilitating the comment of public and private organizations and indviduals by announcing the availability of draft environmental statements and by making copies available to organizations and individuals that request an opportunity to comment. Agencies should devise methods for publicizing the existence of draft statements, for example, by publication of notices in local newspapers or by maintaining a list of groups, including relevant conservation commissions, known to be interested in the agency's activities and directly notifying such groups of the existence of a draft statement, or sending them a copy, as soon as it has been prepared. A copy of the draft statement should in all cases be sent to any applicant whose project is the subject of the statement. Materials to be made available to the public shall be provided without charge to the extent practicable, or at a fee which is not more than the actual cost of reproducing copies required to be sent to other Federal agencies, including the Council.

(e) *Responsibilities of commenting entities:* (1) Agencies and members of the public submitting comments on proposed actions on the basis of draft environmental statements should endeavor to make their comments as specific, substantive, and factual as possible without undue attention to matters of form in the impact statement. Although the comments need not conform to any particular format, it would assist agencies reviewing comments if the comments were organized in a manner consistent with the structure of the draft statement. Emphasis should be placed on the assessment of the environmental impacts of the proposed action, and the acceptability of those impacts on the quality of the environment, particularly as contrasted with the impacts of reasonable alternatives to the action. Commenting entities may recommend modifications to the proposed action and/or new alternatives that will enhance environmental quality and avoid or minimize adverse environmental impacts.

(2) Commenting agencies should indicate whether any of their projects not identified in the draft statement are sufficiently advanced in planning and related environmentally to the proposed action so that a discussion of the environmental interrelationships should be included in the final statement (see § 1500.8(a)(1)). The Council is available to assist agencies in making such determinations.

(3) Agencies and members of the public should indicate in their comments the nature of any monitoring of the environmental effects of the proposed project that appears particularly appropriate. Such monitoring may be necessary during the construction, startup, or operation phases of the project. Agencies with special expertise with respect to the environmental impacts involved are encouraged to assist the sponsoring agency in the establishment and operation of appropriate environmental monitoring.

(f) Agencies seeking comment shall establish time limits of not less than forty-five (45) days for reply, after which it may be presumed, unless the agency or party consulted requests a specified extension of time, that the agency or party consulted has no comment to make. Agencies seeking comment should endeavor to comply with requests for extensions of time of up to fifteen (15) days. In determining an appropriate period for comment, agencies should consider the magnitude and complexity of the statement and the extent of citizen interest in the proposed action.

§ 1500.10   PREPARATION AND CIRCULATION OF FINAL ENVIRONMENTAL STATEMENTS

(a) Agencies should make every effort to discover and discuss all major points of view on the environmental effects of the proposed action and its alternatives in the draft statement itself. However, where opposing professional views and responsible opinion have been overlooked in the draft statement and are brought to the agency's attention through the commenting process, the agency should review the environmental effects of the action in light of those views and should make a meaningful reference in the final statement to the existence of any responsible opposing view not adequately discussed in the draft statement, indicating the agency's response to the issues raised. All substantive comments received on the draft (or summaries thereof where response has been exceptionally voluminous) should be attached to the final statement, whether or not each such comment is thought to merit individual discussion by the agency in the text of the statement.

(b) Copies of final statements, with comments attached, shall be sent to all Federal, State, and local agencies and private organizations that made substantive comments on the draft statement and to individuals who requested a copy of the final statement, as well as any applicant whose project is the subject of the statement. Copies of final statements shall in all cases be sent to the Environmental Protection Agency to assist it in carrying out its responsibilities under section 309 of the Clean Air Act. Where the number of comments on a draft statement is such that distribution of the final statement to all commenting entities appears impracticable, the agency shall consult with the Council concerning alternative arrangements for distribution of the statement.

§ 1500.11   TRANSMITTAL OF STATEMENTS TO THE COUNCIL; MINIMUM PERIODS FOR REVIEW; REQUESTS BY THE COUNCIL

(a) As soon as they have been prepared, ten (10) copies of draft environmental statements, five (5) copies of all comments made thereon (to be forwarded to the Council by the entity making comment at the time comment is forwarded to the responsible agency), and ten (10) copies of the final text of environmental statements (together with the substance of all comments received by the responsible agency from Federal, State, and local agencies and from private organizations and individuals) shall be supplied to the Council. This will serve to meet the statutory requirement to make environmental statements available to the President. At the same time that copies of draft and final statements are sent to the Council, copies should also be sent to relevant commenting entities as set forth in §§ 1500.9 and 1500.10(b) of these guidelines.

(b) To the maximum extent practicable no administrative action subject to section 102(2)(C) is to be taken sooner than ninety (90) days after a draft environmental statement has been circulated for comment, furnished to the Council and, except where advance public disclosure will result in significantly increased costs of procurement to the Government, made available to the public pursuant to these guidelines; neither should such administrative action be taken sooner than thirty (30) days after the final text of an environmental statement (together with comments) has been made available to the Council, commenting agencies, and the public. In all cases, agencies should allot a sufficient review period for the final statement so as to comply with the statutory requirement that the "statement and the comments and views of appropriate Federal, State, and local agencies * * * accompany the proposal through the existing agency review processes." If the final text of an environmental statement is filed within ninety (90) days after a draft statement has been circulated for comment, furnished to the Council and made public pursuant to this section of these guidelines, the minimum thirty (30) day period and the ninety (90) day period may run concurrently to the extent that they overlap. An agency may at any time supplement or amend a draft or final environmental statement, particularly when substantial changes are made in the proposed action, or significant new information becomes available concerning its environmental aspects. In such cases the agency should consult with the Council with respect to the possible need or desirability of recirculation of the statement for the appropriate period.

(c) The Council will publish weekly in the FEDERAL REGISTER lists of environmental statements received during the preceding week that are available for public comment. The date of publication of such lists shall be the date from which the minimum periods for review and advance availability of statements shall be calculated.

(d) The Council's publication of notice of the availability of statements is in addition to the agency's responsibility, as described in § 1500.9(d) of these guidelines, to insure the fullest practicable provision of timely public information concerning the existence and availability of environmental statements. The agency responsible for the environmental statement is also responsible for making the statement, the comments received, and any underlying documents available to the public pursuant to the provisions of the Freedom of Information Act (5 U.S.C. 552), without regard to the exclusion of intra- or interagency memoranda when such memoranda transmit comments of Federal agencies on the environmental impact of the proposed action pursuant to § 1500.9 of these guidelines. Agency procedures prepared pursuant to § 1500.3 (a) of these guidelines shall implement these public information requirements and shall include arrangements for availability of environmental statements and comments at the head and appropriate regional offices of the responsible agency and at appropriate State and areawide clearinghouses unless the Governor of the State involved designates to the Council some other point for receipt of this information. Notice of such designation of an alternate point for receipt of this information will be included in the Office of Management and Budget listing of clearinghouses referred to in § 1500.9(c).

(e) Where emergency circumstances make it necessary to take an action with significant environmental impact without observing the provisions of these guidelines concerning minimum periods for agency review and advance availability of environmental statements, the Federal agency proposing to take the action should consult with the Council about alternative arrangements. Similarly where there are overriding considerations of expense to the Government or impaired program effectiveness, the responsible agency should consult with the Council concerning appropriate modifications of the minimum periods.

(f) In order to assist the Council in fulfilling its responsibilities under the Act and under Executive Order 11514, all agencies shall (as required by section 102(2)(H) of the Act and section 3(i) of Executive Order 11514) be resposive to requests by the council for reports and other information dealing with issues arising in connection with the implementaton of the Act. In particular, agencies shall be responsive to a request by the Council for the preparation and circulation of an environmental statement, unless the agency determines that such a statement is not required, in which case the agency shall prepare an environmental assessment and a publicly available record briefly setting forth the reasons for its determination. In no case, however, shall the Council's silence or failure to comment or request preparation, modification, or recirculation of an environmental statement or to take other action with respect to an environmental statement be construed as bearing in any way on the question of the legal requirement for or the adequacy of such statement under the Act.

§ 1500.12 LEGISLATIVE ACTIONS

(a) The Council and the Office of Management and Budget will cooperate in giving guidance as needed to assist agencies in identifying legislative items believed to have environmental significance. Agencies should prepare impact statements prior to submission of their legislative proposals to the Office of Management and Budget. In this regard, agencies should identify types of repetitive legislation requiring environmental impact statements (such as certain types of bills affecting transportation policy or annual construction authorizations).

(b) With respect to recommendations or reports on proposals for legislation to which section 102(2)(C) applies, the final text of the environmental statement and comments thereon should be available to the Congress and to the public for consideration in connection with the proposed legislation or report. In cases where the scheduling of congressional hearings on recommendations or reports on proposals for legislation which the Federal agency has forwarded to the Congress does not allow adequate time for the completion of a final text of an environmental statement (together with comments), a draft environmental statement may be furnished to the Congress and made available to the public pending transmittal of the comments as received and the final text.

§ 1500.13   APPLICATION OF SECTION 102(2)(C) PROCEDURE TO EXISTING PROJECTS AND PROGRAMS

Agencies have an obligation to reassess ongoing projects and programs in order to avoid or minimize adverse environmental effects. The section 102(2) (C) procedure shall be applied to further major Federal actions having a significant effect on the environment even though they arise from projects or programs initiated prior to enactment of the Act on January 1, 1970. While the status of the work and degree of completion may be considered in determining whether to proceed with the project, it is essential that the environmental impacts of proceeding be reassessed pursuant to the Act's policies and procedures and, if the project or program is continued, that further incremental major actions be shaped so as to enhance and restore environmental quality as well as to avoid or minimize adverse environmental consequences. It is also important in further action that account be taken of environmental consequences not fully evaluated at the outset of the project or program.

§ 1500.14   SUPPLEMENTARY GUIDELINES; EVALUATION OF PROCEDURES

(a) The Council after examining environmental statements and agency procedures with respect to such statements will issue such supplements to these guidelines as are necessary.

(b) Agencies will continue to assess their experience in the implementation of the section 102(2)(C). provisions of the Act and in conforming with these guidelines and report thereon to the Council by June 30, 1974. Such reports should include an identification of the problem areas and suggestions for revision or clarification of these guidelines to achieve effective coordination of views on environmental aspects (and alternatives, where appropriate) of proposed actions without imposing unproductive administrative procedures. Such reports shall also indicate what progress the agency has made in developing substantial criteria and guidance for making environmental assessments as required by § 1500.6(c) of this directive and by section 102(2)(B) of the Act.

Effective date. The revisions of these guidelines shall apply to all draft and final impact statements filed with the Council after January 28, 1974.

Russell E. Train,
Chairman

APPENDIX I   SUMMARY TO ACCOMPANY DRAFT AND FINAL STATEMENTS

(Check one) ( ) Draft. ( ) Final Environmental Statement.

Name of responsible Federal agency (with name of operating division where appropriate). Name, address, and telephone number of individual at the agency who can be contacted for additional information about the proposed action or the statement.

1. Name of action (Check one) ( ) Administrative Action. ( ) Legislative Action.

2. Brief description of action and its purpose. Indicate what States (and counties) particularly affected, and what other proposed Federal actions in the area, if any, are discussed in the statement.

3. Summary of environmental impacts and adverse environmental effects.

4. Summary of major alternatives considered.

5. (For draft statements) List all Federal, State, and local agencies and other parties from which comments have been requested. (For final statements) List all Federal, State, and local agencies and other parties from which written comments have been received.

6. Date draft statement (and final environmental statement, if one has been issued) made available to the Council and the public.

APPENDIX II   AREAS OF ENVIRONMENTAL IMPACT AND FEDERAL AGENCIES AND FEDERAL–STATE AGENCIES [1] WITH JURISDICTION BY LAW OR SPECIAL EXPERTISE TO COMMENT THEREON [2]

AIR

Air Quality

Department of Agriculture—
  Forest Service (effects on vegetation)
Atomic Energy Commission (radioactive substances)
Department of Health, Education, and Welfare
Environmental Protection Agency
Department of the Interior—
  Bureau of Mines (fossil and gaseous fuel combustion)
  Bureau of Sport Fisheries and Wildlife (effect on wildlife)
  Bureau of Outdoor Recreation (effects on recreation)
  Bureau of Land Management (public lands)
  Bureau of Indian affairs (Indian lands)
National Aeronautics and Space Administration (remote sensing, aircraft emissions)
Department of Transportation—
  Assistant Secretary for Systems Development and Technology (auto emissions)
  Coast Guard (vessel emissions)
  Federal Aviation Administration (aircraft emissions)

---

[1] River Basin Commissions (Delaware, Great Lakes, Missouri, New England, Ohio, Pacific Northwest, Souris-Red-Rainy, Susquehanna, Upper Mississippi) and similar Federal-State agencies should be consulted on actions affecting the environment of their specific geographic jurisdictions.

[2] In all cases where a proposed action will have significant international environmental effects, the Department of State should be consulted, and should be sent a copy of any draft and final impact statement which covers such action.

Weather Modification

Department of Agriculture—
  Forest Service
Department of Commerce—
  National Oceanic and Atmospheric Administration
Department of Defense—
  Department of the Air Force
Department of the Interior—
  Bureau of Reclamation
Water Resources Council

WATER

Water Quality

Department of Agriculture—
  Soil Conservation Service
  Forest Service
Atomic Energy Commission (radioactive substances)
Department of the Interior—
  Bureau of Reclamation
  Bureau of Land Management (public lands)
  Bureau of Indian Affairs (Indian lands)
  Bureau of Sport Fisheries and Wildlife
  Bureau of Outdoor Recreation
  Geological Survey
  Office of Saline Water
Environmental Protection Agency
Department of Health, Education, and Welfare
Department of Defense—
  Army Corps of Engineers
  Department of the Navy (ship pollution control)
National Aeronautics and Space Administration (remote sensing)
Department of Transportation—
  Coast Guard (oil spills, ship sanitation)
Department of Commerce—
  National Oceanic and Atmospheric Administration
Water Resources Council
River Basin Commissions (as geographically appropriate)

Marine Pollution, Commercial Fishery Conservation, and Shellfish Sanitation

Department of Commerce—
  National Oceanic and Atmospheric Administration
Department of Defense—
  Army Corps of Engineers
  Office of the Oceanographer of the Navy
Department of Health, Education, and Welfare
Department of the Interior—
  Bureau of Sport Fisheries and Wildlife
  Bureau of Outdoor Recreation
  Bureau of Land Management (outer continental shelf)
  Geological Survey (outer continental shelf)
Department of Transportation—
  Coast Guard
Environmental Protection Agency
National Aeronautics and Space Administration (remote sensing)
Water Resources Council
River Basin Commissions (as geographically appropriate)

Waterway Regulation and Stream Modification

Department of Agriculture—
  Soil Conservation Service
Department of Defense—
  Army Corps of Engineers
Department of the Interior—
  Bureau of Reclamation
  Bureau of Outdoor Recreation
  Fish and Wildlife Service
  Geological Survey
Department of Transportation—
  Coast Guard
Environmental Protection Agency
National Aeronautics and Space Administration (remote sensing)
Water Resources Council
River Basin Commissions (as geographically appropriate)

FISH AND WILDLIFE

Department of Agriculture—
  Forest Service
  Soil Conservation Service
Department of Commerce—
  National Oceanic and Atmospheric Administration (marine species)
Department of the Interior—
  Bureau of Land Management
  Bureau of Outdoor Recreation
  Fish and Wildlife Service
Environmental Protection Agency

SOLID WASTE

Atomic Energy Commission (radioactive waste)
Department of Defense—
  Army Corps of Engineers
Department of Health, Education, and Welfare
Department of the Interior—
  Bureau of Mines (mineral waste, mine acid waste, municipal solid waste, recycling)
  Bureau of Land Management (public lands)
  Bureau of Indian Affairs (Indian lands)
  Geological Survey (geologic and hydrologic effects)
  Office of Saline Water (demineralization)
Department of Transportation—
  Coast Guard (ship sanitation)
Environmental Protection Agency
River Basin Commissions (as geographically appropriate)
Water Resources Council

NOISE

Department of Commerce—
  National Bureau of Standards
Department of Health, Education, and Welfare
Department of Housing and Urban Development (land use and building materials aspects)
Department of Labor—
  Occupational Safety and Health Administration

Department of Transportation—
  Assistant Secretary for Systems Development and Technology
  Federal Aviation Administration, Office of Noise Abatement
Environmental Protection Agency
National Aeronautics and Space Administration

RADIATION

Atomic Energy Commission
Department of Commerce—
  National Bureau of Standards
Department of Health, Education, and Welfare
Department of the Interior—
  Bureau of Mines (uranium mines)
  Mining Enforcement and Safety Administration (uranium mines)
Environmental Protection Agency

HAZARDOUS SUBSTANCES

Toxic Materials

Atomic Energy Commission (radioactive substances)
Department of Agriculture—
  Agricultural Research Service
  Consumer and Marketing Service
Department of Commerce—
  National Oceanic and Atmospheric Administration
Department of Defense
Department of Health, Education, and Welfare
Environmental Protection Agency

Food Additives and Contamination of Foodstuffs

Department of Agriculture—
  Consumer and Marketing Service (meat and poultry products)
Department of Health, Education, and Welfare
Environmental Protection Agency

Pesticides

Department of Agriculture—
  Agricultural Research Service (biological controls, food and fiber production)
  Consumer and Marketing Service
  Forest Service
Department of Commerce—
  National Oceanic and Atmospheric Administration
Department of Health, Education, and Welfare
Department of the Interior—
  Bureau of Sport Fisheries and Wildlife (fish and wildlife effects)
  Bureau of Land Management (public lands)
  Bureau of Indian Affairs (Indian lands)
  Bureau of Reclamation (irrigated lands)
  Fish and Wildlife Service (fish and wildlife effects)
Environmental Protection Agency

Transportation and Handling of Hazardous Materials

Atomic Energy Commission (radioactive substances)
Department of Commerce—
  Maritime Administration
  National Oceanic and Atmospheric Administration (effects on marine life
  and the coastal zone)

Department of Defense—
   Armed Services Explosive Safety Board
   Army Corps of Engineers (navigable waterways)
Department of Transportation—
   Federal Highway Administration, Bureau of Motor Carrier Safety
   Coast Guard
   Federal Railroad Administration
   Federal Aviation Administration
   Assistant Secretary for Systems Development and Technology
   Office of Hazardous Materials
   Office of Pipeline Safety
Environmental Protection Agency

ENERGY SUPPLY AND NATURAL RESOURCES DEVELOPMENT

Electric Energy Development, Generation, Transmission, and Use

Atomic Energy Commission (nuclear)
Department of Agriculture—
   Rural Electrification Administration (rural areas)
Department of Defense—
   Army Corps of Engineers (hydro)
Department of Health, Education, and Welfare (radiation effects)
Department of Housing and Urban Development (urban areas)
Department of the Interior—
   Bureau of Indian Affairs (Indian lands)
   Bureau of Land Management (public lands)
   Bureau of Reclamation
   Power Marketing Administration
   Geological Survey
   Bureau of Outdoor Recreation
   Fish and Wildlife Service
   National Park Service
Environmental Protection Agency
Federal Power Commission (hydro, transmission, and supply)
River Basin Commissions (as geographically appropriate)
Tennessee Valley Authority
Water Resources Council

Petroleum Development, Extraction, Refining, Transport, and Use

Department of the Interior—
   Office of Oil and Gas
   Bureau of Mines
   Geological Survey
   Bureau of Land Management (public lands and outer continental shelf)
   Bureau of Indian Affairs (Indian lands)
   Fish and Wildlife Service (effects on fish and wildlife)
   Bureau of Outdoor Recreation
   National Park Service
Department of Transportation (Transport and Pipeline Safety)
Environmental Protection Agency
Interstate Commerce Commission

Natural Gas Development, Production, Transmission, and Use

Department of Housing and Urban Development (urban areas)
Department of the Interior—
   Office of Oil and Gas
   Geological Survey

Department of the Interior—Continued
  Bureau of Mines
  Bureau of Land Management (public lands)
  Bureau of Indian Affairs (Indian lands)
  Bureau of Outdoor Recreation
  Fish and Wildlife Service
  National Park Service
Department of Transportation (transport and safety)
Environmental Protection Agency
Federal Power Commission (production, transmission, and supply)
Interstate Commerce Commission

Coal and Minerals Development, Mining, Conversion, Processing, Transport, and Use

Appalachian Regional Commission
Department of Agriculture—
  Forest Service
Department of Commerce
Department of the Interior—
  Office of Coal Research
  Mining Enforcement and Safety Administration
  Bureau of Mines
  Geological Survey
  Bureau of Indian Affairs (Indian lands)
  Bureau of Land Management (public lands)
  Bureau of Outdoor Recreation
  Fish and Wildlife Service
  National Park Service
Department of Labor—
  Occupational Safety and Health Administration
Department of Transportation
Environmental Protection Agency
Interstate Commerce Commission
Tennessee Valley Authority

Renewable Resource Development, Production, Management, Harvest, Transport, and Use

Department of Agriculture—
  Forest Service
  Soil Conservation Service
Department of Commerce
Department of Housing and Urban Development (building materials)
Department of the Interior—
  Geological Survey
  Bureau of Land Management (public lands)
  Bureau of Indian Affairs (Indian lands)
  Bureau of Outdoor Recreation
  Fish and Wildlife Service
  National Park Service
Department of Transportation
Environmental Protection Agency
Interstate Commerce Commission (freight rates)

Energy and Natural Resources Conservation

Department of Agriculture—
  Forest Service
  Soil Conservation Service

Department of Commerce—
    National Bureau of Standards (energy efficiency)
Department of Housing and Urban Development—
    Federal Housing Administration (housing standards)
Department of the Interior—
    Office of Energy Conservation
    Bureau of Mines
    Bureau of Reclamation
    Geological Survey
    Power Marketing Administration
Department of Transportation
Environmental Protection Agency
Federal Power Commission
General Services Administration (design and operation of buildings)
Tennessee Valley Authority

LAND USE AND MANAGEMENT

Land Use Changes, Planning and Regulation of Land Development

Department of Agriculture—
    Forest Service (forest lands)
    Agricultural Research Service (agricultural lands)
Department of Housing and Urban Development
Department of the Interior—
    Office of Land Use and Water Planning
    Bureau of Land Management (public lands)
    Bureau of Indian Affairs (Indian lands)
    Bureau of Sport Fisheries and Wildlife (wildlife refuges)
    Bureau of Outdoor Recreation (recreation lands)
    Fish and Wildlife Service
    National Park Service (NPS units)
Department of Transportation
Environmental Protection Agency (pollution effects)
National Aeronautics and Space Administration (remote sensing)
River Basin Commissions (as geographically appropriate)

Public Land Management

Department of Agriculture—
    Forest Service (forests)
Department of Defense
Department of the Interior—
    Bureau of Land Management
    Bureau of Indian Affairs (Indian lands)
    Bureau of Outdoor Recreation (recreation lands)
    Fish and Wildlife Service (wildlife refuges)
    National Park Service (NPS units)
Federal Power Commission (project lands)
General Services Administration
National Aeronautics and Space Administration (remote sensing)
Tennessee Valley Authority (project lands)

Protection of Environmentally Critical Areas—Floodplains, Wetlands, Beaches and Dunes, Unstable Soils, Steep Slopes, Aquifer Recharge Areas, etc.

Department of Agriculture—
    Agricultural Stabilization and Conservation Service
    Soil Conservation Service
    Forest Service

Department of Commerce—
    National Oceanic and Atmospheric Administration (coastal areas)
Department of Defense—
    Army Corps of Engineers
Department of Housing and Urban Development (urban and floodplain areas)
Department of the Interior—
    Office of Land Use and Water Planning
    Bureau of Outdoor Recreation
    Bureau of Reclamation
    Bureau of Land Management
    Fish and Wildlife Service
    Geological Survey
Environmental Protection Agency (pollution effects)
National Aeronautics and Space Administration (remote sensing)
River Basins Commissions (as geographically appropriate)
Water Resources Council

Land Use in Coastal Areas

Department of Agriculture—
    Forest Service
    Soil Conservation Service (soil stability, hydrology)
Department of Commerce—
    National Oceanic and Atmospheric Administration (impact on marine life and coastal zone management)
Department of Defense—
    Army Corps of Engineers (beaches, dredge and fill permits, Refuse Act permits)
Department of Housing and Urban Development (urban areas)
Department of the Interior—
    Office of Land Use and Water Planning
    Fish and Wildlife Service
    National Park Service
    Geological Survey
    Bureau of Outdoor Recreation
    Bureau of Land Management (public lands)
Department of Transportation—
    Coast Guard (bridges, navigation)
Environmental Protection Agency (pollution effects)
National Aeronautics and Space Administration (remote sensing)

Redevelopment and Construction in Built-Up Areas

Department of Commerce—
    Economic Development Administration (designated areas)
Department of Housing and Urban Development
Department of the Interior—
    Office of Land Use and Water Planning
Department of Transportation
Environmental Protection Agency
General Services Administration
Office of Economic Opportunity

Density and Congestion Mitigation

Department of Health, Education, and Welfare
Department of Housing and Urban Development
Department of the Interior—
    Office of Land Use and Water Planning
    Bureau of Outdoor Recreation
Department of Transportation
Environmental Protection Agency

Neighborhood Character and Continuity

Department of Health, Education, and Welfare
Department of Housing and Urban Development
National Endowment for the Arts
Office of Economic Opportunity

Impacts on Low-Income Populations

Department of Commerce—
    Economic Development Administration (designated areas)
Department of Health, Education, and Welfare
Department of Housing and Urban Development
Office of Economic Opportunity

Historic, Architectural, and Archeological Preservation

Advisory Council on Historic Preservation
Department of Housing and Urban Development
Department of the Interior—
    National Park Service
    Bureau of Land Management (public lands)
    Bureau of Indian Affairs (Indian lands)
General Services Administration
National Endowment for the Arts

Soil and Plant Conservation and Hydrology

Department of Agriculture—
    Soil Conservation Service
    Agricultural Service
    Forest Service
Department of Commerce—
    National Oceanic and Atmospheric Administration
Department of Defense—
    Army Corps of Engineers (dredging, aquatic plants)
Department of Health, Education, and Welfare
Department of the Interior—
    Bureau of Land Management
    Fish and Wildlife Service
    Geological Survey
    Bureau of Reclamation
Environmental Protection Agency
National Aeronautics and Space Administration (remote sensing)
River Basin Commissions (as geographically appropriate)
Water Resources Council

Outdoor Recreation

Department of Agriculture—
    Forest Service
    Soil Conservation Service
Department of Defense—
    Army Corps of Engineers
Department of Housing and Urban Development (urban areas)
Department of the Interior—
    Bureau of Land Management
    National Park Service
    Bureau of Outdoor Recreation
    Bureau of Indian Affairs
    Fish and Wildlife Service
Environmental Protection Agency

National Aeronautics and Space Administration (remote sensing)
River Basin Commissions (as geographically appropriate)
Water Resources Council

APPENDIX III   OFFICES WITHIN FEDERAL AGENCIES AND FEDERAL–STATE AGENCIES FOR INFORMATION REGARDING THE AGENCIES' NEPA ACTIVITIES AND FOR RECEIVING OTHER AGENCIES' IMPACT STATEMENTS FOR WHICH COMMENTS ARE REQUESTED

ADVISORY COUNCIL ON HISTORIC PRESERVATION

Office of Review and Compliance, Advisory Council on Historic Preservation,
    1522 K Street, N.W., Suite 430, Washington, D.C. 20005   (202)
    254–3380

DEPARTMENT OF AGRICULTURE [1]

Office of the Secretary, Attn: Coordinator Environmental Quality Activities,
    U.S. Department of Agriculture, Washington, D.C. 20250   (202)
    447–3965

APPALACHIAN REGIONAL COMMISSION

Office of the Alternate Federal Co-Chairman, Appalachian Regional Com-
    mission, 1666 Connecticut Avenue, N.W., Washington, D.C. 20235
    (202)   967–4103

DEPARTMENT OF THE ARMY (CORPS OF ENGINEERS)

Executive Director of Civil Works, Office of the Chief of Engineers, U.S.
    Army Corps of Engineers, Washington, D.C. 20314   (202)   693–7093

DEPARTMENT OF COMMERCE

Office of the Deputy Assistant Secretary for Environmental Affairs, U.S.
    Department of Commerce, Washington, D.C. 20230   (202)   967–4335

DEPARTMENT OF DEFENSE

Office of the Assistant to the Secretary of Defense for Environmental Quality,
    U.S. Department of Defense, Room 3D171, The Pentagon, Washington,
    D.C. 20301   (202)   695–3010

DELAWARE RIVER BASIN COMMISSION

Environmental Unit, Delaware River Basin Commission, Post Office Box 360,
    Trenton, N.J. 08603   (609)   883–9500

---

[1] Requests for comments or information from individual units of the Department of Agriculture, e.g., Soil Conservation Service, Forest Service, etc.
should be sent to the Office of the Secretary, Department of Agriculture, at
the address given above.

ENERGY RESEARCH AND DEVELOPMENT ADMINISTRATION

Office of the Assistant Administrator for Environment and Safety, Energy Research and Development Administration, Washington, D.C. 20545 (202) 937–3208

ENVIRONMENTAL PROTECTION AGENCY [2]

Director, Office of Federal Activities, Environmental Protection Agency, 401 M Street, S.W., Washington, D.C. 20460   (202)   755–0777

FEDERAL POWER COMMISSION

Office of Energy Systems, Federal Power Commission, Room 4306, 825 North Capitol Street, N.E., Washington, D.C. 20426   (202)   386–3335

---

[2] Contact the Office of Federal Activities for environmental statements concerning legislation, regulations, national program proposals or other major policy issues.

For all other EPA consultation, contact the Regional Administrator in whose area the proposed action (e.g., highway or water resource construction projects) will take place. The Regional Administrators will coordinate the EPA review. Addresses of the Regional Administrators and the areas covered by their regions are as follows:

Regional Administrator I,
U.S. Environmental Protection Agency
Room 2303, John F. Kennedy Federal Bldg.
Boston, Mass. 02203
(617)   223–4635

Connecticut, Maine, Massachusetts, New Hampshire, Rhode Island, Vermont

Regional Administrator II,
U.S. Environmental Protection Agency
Room 908, 26 Federal Plaza
New York, N.Y. 10007
(212)   264–8556

New Jersey, New York, Puerto Rico, Virgin Islands

Regional Administrator III,
U.S. Environmental Protection Agency
Curtis Bldg., 6th & Walnut Streets
Philadelphia, Pa. 19106
(215)   597–8332

Delaware, Maryland, Pennsylvania, Virginia, West Virginia, District of Columbia

Regional Administrator IV,
U.S. Environmental Protection Agency
1421 Peachtree Street, N.E.
Atlanta, Ga.   30309
(404)   526–5458

Alabama, Florida, Georgia, Kentucky, Mississippi, North Carolina, South Carolina, Tennessee

Regional Administrator V,
U.S. Environmental Protection Agency
230 South Dearborn Street
Chicago, Ill. 60604
(312)   353–5758

Illinois, Indiana, Michigan, Minnesota, Ohio, Wisconsin

(Continued)

GENERAL SERVICES ADMINISTRATION

Office of Environmental Affairs, Office of the Deputy Administrator for Special Projects, General Services Administration, Washington, D.C. 20405 (202) 343–4161

GREAT LAKES BASIN COMMISSION

Office of the Chairman, Great Lakes Basin Commission, 3475 Plymouth Road, P.O. Box 999, Ann Arbor, Michigan 48106   (313)   769–7431

DEPARTMENT OF HEALTH, EDUCATION, AND WELFARE [3]

Office of Environmental Affairs, Office of the Assistant Secretary for Administration and Management, Department of Health, Education, and Welfare, 330 Independence Avenue, S.W., Washington, D.C. 20201   (202) 245–7243

(Continued)

Regional Administrator VI, U.S. Environmental Protection Agency 1600 Patterson Street Suite 1100 Dallas, Tex. 75201 (214) 749–1237	Arkansas, Louisiana, New Mexico, Texas, Oklahoma
Regional Administrator VII, U.S. Environmental Protection Agency 1735 Baltimore Avenue Kansas City, Mo. 64108 (816) 374–2921	Iowa, Kansas, Missouri, Nebraska
Regional Administrator VIII, U.S. Environmental Protection Agency Suite 900, Lincoln Tower 1860 Lincoln Street Denver, Colo. 80203 (303) 837–4831	Colorado, Montana, North Dakota, South Dakota, Utah, Wyoming
Regional Administrator IX, U.S. Environmental Protection Agency 100 California Street San Francisco, Calif. 94111 (415) 556–0970	Arizona, California, Hawaii, Nevada, American Samoa, Guam, Trust Territories of Pacific Islands, Wake Island
Regional Administrator X, U.S. Environmental Protection Agency 1200 Sixth Avenue Seattle, Wash. 98101 (206) 442–1595	Alaska, Idaho, Oregon, Washington

[3] Contact the Office of Environmental Affairs for information on HEW's environmental statements concerning legislation, regulations, national program proposals or other major policy issues, and for all requests for HEW comments on impact statements of other agencies.

(Continued)

DEPARTMENT OF HOUSING AND URBAN DEVELOPMENT [4]

Director, Office of Environmental Quality, Department of Housing and Urban Development, Room 7258, Washington, D.C. 20410   (202) 755–6309

(Continued)

For information with respect to HEW actions occurring within the jurisdiction of the Departments' Regional Directors, contact the appropriate Regional Environmental Officer:

Region I:

Regional Environmental Officer, U.S. Department of Health, Education, and Welfare, John F. Kennedy Center, Room E–311, Boston, Mass. 02203 (617)   223–5485

Region II:

Regional Environmental Officer, U.S. Department of Health, Education, and Welfare, 26 Federal Plaza, Room 3309, New York, N.Y. 10007 (212)   264–4483

Region III:

Regional Environmental Officer, U.S. Department of Health, Education, and Welfare, P.O. Box 13716, Philadelphia, Pa. 19101   (215)   597–6476

Region IV:

Regional Environmental Officer, U.S. Department of Health, Education, and Welfare, 50 Seventh Street, N.E., Room 447, Atlanta, Ga. 30323 (404)   526–5048

Region V:

Regional Environmental Officer, U.S. Department of Health, Education, and Welfare, 300 South Wacker Drive, 35th floor, Chicago, Ill. 60606 (312)   353–1670

Region VI:

Regional Environmental Officer, U.S. Department of Health, Education, and Welfare, 1114 Commerce Street, Room 925, Dallas, Tex. 75202 (214)   749–7741

Region VII:

Regional Environmental Officer, U.S. Department of Health, Education, and Welfare, 601 East 12th Street, Kansas City, Mo. 64106   (816) 374–5912

Region VIII:

Regional Environmental Officer, U.S. Department of Health, Education, and Welfare, 1916 Stout Street, 10th floor, Denver, Colo. 80202   (303) 837–2875

Region IX:

Regional Environmental Officer, U.S. Department of Health, Education, and Welfare, 50 Fulton Street, Room 410, San Francisco, Calif. 94102 (415)   556–3687

Region X:

Regional Environmental Officer, U.S. Department of Health, Education, and Welfare, 1321 Second Street, Mail Stop 610, Seattle, Wash. 98101 (206)   442–7790

[4] Contact the Director with regard to environmental impacts of legislation, policy statements, program regulations and procedures, and precedent-making project decisions. For all other consultation, contact the HUD Regional Administrator in whose jurisdiction the project lies, as follows:

Regional Administrator I, Environmental Clearance Officer, U.S. Department of Housing and Urban Development, John F. Kennedy Federal Building, Room 405, Boston, Mass. 02203   (617)   223–4347

(Continued)

DEPARTMENT OF THE INTERIOR [5]

Director, Office of Environmental Project Review, Department of the Interior, Interior Building, Washington, D.C. 20240   (202) 343-3891

INTERSTATE COMMERCE COMMISSION

Office of Proceedings, Environmental Section, Interstate Commerce Commission, Washington, D.C. 20423   (202) 343-7966

DEPARTMENT OF LABOR

Office of Manpower for Policy Evaluation and Research, Division of Manpower Policy and Planning, Department of Labor, 601 D Street, N.W., Washington, D.C. 20210   (202) 376-6274

MISSOURI RIVER BASIN COMMISSION

Office of the Chairman, Missouri River Basin Commission, 10050 Regency Circle, Omaha, Nebr. 68114   (402) 397-5714

NATIONAL AERONAUTICS AND SPACE ADMINISTRATION

Office of Policy Analysis, National Aeronautics and Space Administration, 400 Maryland Avenue, Washington, D.C. 20546   (202) 755-8433

(Continued)

Regional Administrator II, Environmental Clearance Officer, U.S. Department of Housing and Urban Development, 26 Federal Plaza, New York, N.Y. 10007   (212) 264-4197

Regional Administrator III, Environmental Clearance Officer, U.S. Department of Housing and Urban Development, Curtis Building, Sixth and Walnut Streets, Philadelphia, Pa. 19106   (215) 597-2520

Regional Administrator IV, Environmental Clearance Officer, U.S. Department of Housing and Urban Development, Pershing Point Plaza, 1371 Peachtree Street, N.E., Atlanta, Ga. 30309   (404) 526-3521

Regional Administrator V, Environmental Clearance Officer, U.S. Department of Housing and Urban Development, 300 South Wacker Drive, Chicago, Ill. 60606   (312) 353-1680

Regional Administrator VI, Environmental Clearance Officer, U.S. Department of Housing and Urban Development, U.S. Courthouse, 1100 Commerce Street, Dallas, Tex. 75202   (214) 749-7466

Regional Administrator VII, Environmental Clearance Officer, U.S. Department of Housing nad Urban Development, 911 Walnut Street, Kansas City, Mo. 64106   (816) 374-3191

Regional Administrator VIII, Environmental Clearance Officer, U.S. Department of Housing and Urban Development, 1961 Stout Street, Denver, Colo. 80202   (303) 837-3811

Regional Administrator IX, Environmental Clearance Officer, U.S. Department of Housing and Urban Development, 450 Golden Gate Avenue, P.O. Box 36003, San Francisco, Calif. 94102   (415) 556-5720

Regional Administrator X, Environmental Clearance Officer, U.S. Department of Housing and Urban Development, 1321 Second Avenue, Seattle, Wash. 98101   (206) 442-7790

[5] Requests for comments or information from individual units of the Department of the Interior should be sent to the Office of Environmental Project Review at the address given above.

NATIONAL CAPITAL PLANNING COMMISSION
Office of Environmental Affairs, National Capital Planning Commission, Washington, D.C. 20576   (202)   382–7200

NATIONAL ENDOWMENT FOR THE ARTS
Office of the Director, Architecture and Environmental Arts Program, National Endowment for the Arts, Washington, D.C. 20506   (202)   634–4276

NEW ENGLAND RIVER BASIN COMMISSION
Office of the Chairman, New England River Basin Commission, 55 Court Street, Boston, Mass. 02108   (617)   223–6244

NUCLEAR REGULATORY COMMISSION
For Nuclear Power Plants: Office of the Assistant Director for Environmental Projects, Division of Nuclear Reactor Licensing, Nuclear Regulatory Commission, Washington, D.C. 20555   (202)   443–6965
For Fuel-cycle Projects: Fuel-cycle Environmental Projects Branch, Division of Materials and Fuel-cycle Facility Licensing, Nuclear Regulatory Commission, Washington, D.C. 20555   (202)   492–7631

OHIO RIVER BASIN COMMISSION
Office of the Chairman, Ohio River Basin Commission, 36 East 4th Street, Suite 208, Cincinnati, Ohio 45202   (513)   684–3831

PACIFIC NORTHWEST RIVER BASIN COMMISSION
Office of the Chairman, Pacific Northwest River Basin Commission, Post Office Box 908, Vancouver, Wash. 98660   (206)   694–2581

SOURIS-RED-RAINY RIVER BASINS COMMISSION
Office of the Chairman, Souris-Red-Rainy River Basins Commission, Suite 6, Professional Building, Holiday Mall, Moorhead, Minnesota 56560 (701) 237–5227

DEPARTMENT OF STATE
Office of Environmental Affairs, Room 7822, Department of State, Washington, D.C. 20520   (202)   632–9278

SUSQUEHANNA RIVER BASIN COMMISSION
Office of the Executive Director, Susquehanna River Basin Commission, 5012 Lenker Street, Mechanicsburg, Pa. 17055   (717)   737–0501

TENNESSEE VALLEY AUTHORITY
Office of the Chief of Environmental Assessment and Compliance Staff, Tennessee Valley Authority, 267—401 Building, Chattanooga, Tenn. 37401 (615)   755–3175

DEPARTMENT OF TRANSPORTATION [6]

Director, Office of Environmental Affairs, Office of the Assistant Secretary for Environment, Safety, and Consumer Affairs, Department of Transportation, Washington, D.C. 20590   (202)   426–4357

---

[6] Contact the Office of Environmental Quality, Department of Transportation, for information on DOT's environmental statements concerning legislation, regulations, national program proposals, or other major policy issues.

For information regarding the Department of Transportation's other environmental statements, contact the national office for the appropriate administration:

U.S. Coast Guard
Office of Ports and Waterways Planning, U.S. Coast Guard, 400 Seventh Street, S.W., Washington, D.C. 20590   (202)   426–8722
Federal Aviation Administration
Office of Environmental Quality, Federal Aviation Administration, 800 Independence Avenue, S.W., Washington, D.C. 20591   (202)   426–8722
Federal Highway Administration
Office of Environmental Policy, Environmental Programs Division, Federal Highway Administration, 400 Seventh Street, S.W., Washington, D.C. 20590   (202)   426–0106
Federal Railroad Administration
Office of the Chief Counsel, Federal Railroad Administration, 400 Seventh Street, S.W., Washington, D.C. 20590   (202)   426–0767
Urban Mass Transportation Administration
Capitol Assistance, Urban Mass Transportation Administration, 400 Seventh Street, S.W., Washington, D.C. 20590   (202)   426–1985

For other administrations not listed above, contact the Office of Environmental Quality, Department of Transportation, at the address given above.

For comments on other agencies' environmental statements, contact the appropriate administration's regional office. If more than one administration within the Department of Transportation is to be requested to comment, contact the Secretarial Representative in the appropriate Regional Office for coordination of the Department's comments:
Secretarial Representative
Region I Secretarial Representative, U.S. Department of Transportation, Transportation Systems Center, 55 Broadway, Cambridge, Mass. 02142   (617)   494–2709
Region II Secretarial Representative, U.S. Department of Transportation, 26 Federal Plaza, Room 1811, New York, N.Y. 10007   (212)   264–2672
Region III Secretarial Representative, U.S. Department of Transportation, 434 Walnut Street, 10th floor, Philadelphia, Pa. 19106   (215)   597–9430
Region IV Secretarial Representative, U.S. Department of Transportation, Suite 515, 1720 Peachtree Rd., N.W., Atlanta, Ga. 30309   (404)   526–3738
Region V Secretarial Representative, U.S. Department of Transportation, 17th Floor, 300 S. Wacker Drive, Chicago, Ill. 60606   (312)   353–4000
Region VI Secretarial Representative, U.S. Department of Transportation, 9–C–18 Federal Center, 1100 Commerce Street, Dallas, Tex. 75202   (214)   749–1851
Region VII Secretarial Representative, U.S. Department of Transportation, 601 East 12th Street, Room 634, Kansas City, Mo. 64106   (816)   374–5801
Region VIII Secretarial Representative, U.S. Department of Transportation, Prudential Plaza, Suite 1822, 1050 17th Street, Denver, Colo. 80202   (303)   837–3242

(Continued)

DEPARTMENT OF THE TREASURY

Office of Assistant Secretary for Administration, Department of the Treasury, Washington, D.C. 20220   (202)   964–5391

(Continued)

Region IX Secretarial Representative, U.S. Department of Transportation, 450 Golden Gate Avenue, Box 36133, San Francisco, Calif. 94102 (415)   556–5961

Region X Secretarial Representative, U.S. Department of Transportation, 3112 Federal Building, 915 Second Street, Seattle, Wash. 98174   (206) 442–0590

Federal Aviation Administration

New England Region, Office of the Regional Director, Federal Aviation Administration, 12 Northeast Executive Park, Burlington, Mass. 01803 (617)   467–7244

Eastern Region, Office of the Regional Director, Federal Aviation Administration, Federal Building, JFK International Airport, Jamaica, N.Y. 11430 (212)   995–3333

Southern Region, Office of the Regional Director, Federal Aviation Administration, P.O. Box 20636, Atlanta, Ga. 30320   (404)   526–7222

Great Lakes Region, Office of the Regional Director, Federal Aviation Administration, 2300 East Devon, Des Plaines, Ill. 60018   (312)   694–4500

Southwest Region, Office of the Regional Director, Federal Aviation Administration, P.O. Box 1689, Fort Worth, Tex. 76101   (817)   624–4911

Central Region, Office of the Regional Director, Federal Aviation Administration, 601 East 12th Street, Kansas City, Mo. 64106   (816)   374–5626

Rocky Mountain Region, Office of the Regional Director, Federal Aviation Administration, Park Hill Station, P.O. Box 7213, Denver, Colo. 80207 (303)   837–3646

Western Region, Office of the Regional Director, Federal Aviation Administration, P.O. Box 92007, Worldway Postal Center, Los Angeles, Calif. 90009 (213)   536–6435

Northwest Region, Office of the Regional Director, Federal Aviation Administration, FAA Building, 9010 East Marginal Way South, King County International Airport, Seattle, Wash. 98108   (206)   767–2780

Federal Highway Administration

Region 1:
Regional Administrator, Federal Highway Administration, 4 Normanskill Boulevard, Delmar, N.Y. 12054   (518)   472–6476

Region 3:
Regional Administrator, Federal Highway Administration, George H. Fallon Federal Office Building, Room 1633, 31 Hopkins Plaza, Baltimore, Md. 21201   (301)   962–2361

Region 4:
Regional Administrator, Federal Highway Administration, 1720 Peachtree Road, N.W., Suite 200, Atlanta, Ga. 30309   (404)   526–5078

Region 5:
Regional Administrator, Federal Highway Administration, 18209 Dixie Highway, Homewood, Ill. 60430   (312)   799–6300

Region 6:
Regional Administrator, Federal Highway Administration, 819 Taylor Street, Fort Worth, Tex. 76102   (817)   334–3232

Region 7:
Regional Administrator, Federal Highway Administration, P.O. Box 7186, Country Club Station, Kansas City, Mo. 64113   (816)   926–7563

(Continued)

UPPER MISSISSIPPI RIVER BASIN COMMISSION

Office of the Chairman, Upper Mississippi River Basin Commission, Federal Office Building, Room 510, Fort Snelling, Twin Cities, Minn. 55111 (612) 725-4690

WATER RESOURCES COUNCIL

Office of the Assistant Director, Planning Division, Water Resources Council, 2120 L Street, N.W., Suite 800, Washington, D.C. 20037 (202) 254-6442

(Continued)

Region 8:
Regional Administrator, Federal Highway Administration, Building 40, Room 242, Denver Federal Center, Denver, Colo. 80225 (303) 236-4051
Region 9:
Regional Administrator, Federal Highway Administration, 450 Golden Gate Avenue, Box 36112, San Francisco, Calif. 94102 (415) 556-6415
Region 10:
Regional Administrator, Federal Highway Administration, Mohawk Building, Room 412, 222 Southwest Morrison Street, Portland, Oreg. 97204 (503) 221-2065

Urban Mass Transportation Administration

Region I, Office of the UMTA Representative, Urban Mass Transportation Administration, Transportation Systems, Kendall Square, Cambridge, Mass. 02142 (617) 494-2005
Region II, Office of the UMTA Representative, Urban Mass Transportation Administration, 26 Federal Plaza, Suite 507, New York, N.Y. 10007 (212) 264-8162
Region III, Office of the UMTA Representative, Urban Mass Transportation Administration, 434 Walnut Street, 10th floor, Philadelphia, Pa. 19106 (215) 597-8098
Region IV, Office of UMTA Representative, Urban Mass Transportation Administration, 1720 Peachtree Road, Northwest, Suite 501, Atlanta, Ga. 30309 (404) 526-3948
Region V, Office of the UMTA Representative, Urban Mass Transportation Administration, 300 South Wacker Drive, Suite 700, Chicago, Ill. 60606 (312) 353-6005
Region VI, Office of the UMTA Representative, Urban Mass Transportation Administration, Federal Center, Suite 9A32, 819 Taylor Street, Fort Worth, Tex. 76102 (817) 334-3896
Region VII, Office of the UMTA Representative, Urban Mass Transportation Administration, Suite 633, 601 East 12th Street, Kansas City, Mo. 64106 (816) 374-5845
Region VIII, Office of the UMTA Representative, Urban Mass Transportation Administration, Prudential Plaza, Suite 1822, 1050 17th Street, Denver, Colo. 80202 (303) 837-3242
Region IX, Office of the UMTA Representative, Urban Mass Transportation Administration, 450 Golden Gate Avenue, Box 36125, San Francisco, Calif. 94102 (415) 556-2884
Region X, Office of the UMTA Representative, Urban Mass Transportation Administration, Federal Building, Suite 3106, 915 Second Avenue, Seattle, Wash. 98174 (206) 442-4210

APPENDIX IV   STATE AND LOCAL AGENCY REVIEW OF IMPACT STATEMENTS

1. OMB Circular No. A–95 through its system of clearinghouses provides a means for securing the views of State and local environmental agencies, which can assist in the preparation of impact statements. Under A–95, review of the proposed project in the case of federally assisted projects (Part I of A–95) generally takes place prior to the preparation of the impact statement. Therefore, comments on the environmental effects of the proposed project that are secured during this stage of the A–95 process represent inputs to the environmental impact statement.

2. In the case of direct Federal development (Part II of A–95), Federal agencies are required to consult with clearinghouses at the earliest practicable time in the planning of the project or activity. Where such consultation occurs prior to completion of the draft impact statement, comments relating to the environmental effects of the proposed action would also represent inputs to the environmental impact statement.

3. In either case, whatever comments are made on environmental effects of proposed Federal or federally assisted projects by clearinghouses, or by State and local environmental agencies through clearinghouses, in the course of the A–95 review should be attached to the draft impact statement when it is circulated for review. Copies of the statement should be sent to the agencies making such comments. Whether those agencies then elect to comment again on the basis of the draft impact statement is a matter to be left to the discretion of the commenting agency depending on its resources, the significance of the project, and the extent to which its earlier comments were considered in preparing the draft statement.

4. The clearinghouses may also be used, by mutual agreement, for securing reviews of the draft environmental impact statement. However, the Federal agency may wish to deal directly with appropriate State or local agencies in the review of impact statements because the clearinghouses may be unwilling or unable to handle this phase of the process. In some cases, the Governor may have designated a specific agency, other than the clearinghouse, for securing reviews of impact statements. In any case, the clearinghouses should be sent copies of the impact statement.

5. To aid clearinghouses in coordinating State and local comments, draft statements should include copies of State and local agency comments made earlier under the A–95 process and should indicate on the summary sheet those other agencies from which comments have been requested, as specified in Appendix I of the CEQ Guidelines.

C

Proposed CEQ Regulations to Implement the National Environmental Policy Act

PROPOSED RULES

[3125-01]

COUNCIL ON ENVIRONMENTAL QUALITY

[40 CFR Parts 1500, 1501, 1502, 1503, 1504, 1505, 1506, 1507, 1508]

NATIONAL ENVIRONMENTAL POLICY ACT—REGULATIONS

Proposed Implementation of Procedural Provisions

MAY 31, 1978.

AGENCY: Council on Environmental Quality, Executive Office of the President.

ACTION: Proposed regulations.

SUMMARY: These proposed regulations implementing procedural provisions of the National Environmental Policy Act are submitted for public comment. These regulations would provide Federal agencies with uniform procedures for implementing the law. The regulations would accomplish three principal aims: to reduce paperwork, to reduce delays, and to produce better decisions.

DATES: Comments must be received by August 11, 1978.

ADDRESSES: Comments should be addressed to: Nicholas C. Yost, General Counsel, Attention: NEPA Comments, Council on Environmental Quality, 722 Jackson Place NW., Washington, D.C. 20006.

FOR FURTHER INFORMATION CONTACT:

Nicholas C. Yost, General Council on Environmental Quality (address same as above), 202-633-7032.

SUPPLEMENTARY INFORMATION:

1. PURPOSE

We are publishing for public review draft regulations to implement the National Environmental Policy Act. Their purpose is to provide all Federal agencies with an efficient, uniform procedure for translating the law into practical action. We expect the new regulations to accomplish three principal aims: To reduce paperwork, to reduce delays, and at the same time to produce better decisions, thereby better accomplishing the law's objective, which is to protect and enhance the quality of the human environment.

These regulations replace the Guidelines issued by previous Councils, under Executive Order 11514 (1970), and apply more broadly. The Guidelines assist Federal agencies in carrying out NEPA's most conspicuous requirement, the preparation of environmental impact statements (EISs). These regulations were developed in response to Executive Order 11991 issued by President Carter in 1977, and implement "the procedural provisions of the Act." They address all nine subdivisions of Section 102(2) of the Act, rather than just the EIS provision covered by the Guidelines, and they carry out the broad purposes and spirit of the Act.

President Carter instructed us that the regulations should be:

* * * designed to make the enviromental impact statement more useful to decision-makers and the public; and to reduce paperwork and the accumulation of extraneous background data, in order to emphasize the need to focus on real environmental issues and alternatives.

The President has also signed Executive Order 12044, dealing with regulatory reform. It is our intention that that Order and these NEPA regulations be read together and implemented consistently.

2. SUMMARY OF CHANGES MADE BY THE REGULATIONS

Following this mandate in developing the new regulations, we have kept in mind the threefold objective of less paperwork, less delay, and better decisions.

A. REDUCING PAPERWORK

The measures to reduce paperwork are listed in sec. 1500.4 of the regulations. Neither NEPA nor these regulations impose paperwork requirements on the public. These regulations reduce such requirements on agencies of government.

i. *Reducing the length of environmental impact statements.* Agencies are directed to write concise EISs, which shall normally be less than 150 pages, or, for proposals of unusual scope and complexity, 300 pages.

ii. *Emphasize options among alternatives.* The regulations stress that the environmental analysis is to concentrate on alternatives, which are the heart of the matter; to treat peripheral matters briefly; and to avoid accumulating masses of background data which tend to obscure the important issues.

iii. *Using an early "scoping" process to determine what the important issues are.* To assist agencies in deciding what the central issues are, how long the EIS shall be, and how the responsibility for the EIS will be allocated among the lead agency and cooperating agencies, a new "scoping" procedure is established. Scoping meetings are to be held as early in the NEPA process as possible—in most cases, shortly after the decision to prepare an EIS—and shall be integrated with other planning.

iv. *Writing in plain language.* The regulations strongly advocate writing in plain, direct language.

v. *Following a clear format.* The regulations spell out a standard format intended to eliminate repetitive discussion, stress the major conclusions, highlight the areas of controversy, and focus on the issues to be resolved.

vi. *Requiring summaries of environmental impact statements* to make the document more usable by more people.

vii. *Eliminating duplication.* To eliminate duplication, the regulations provide for Federal agencies to prepare EISs jointly with state and local units of government which have "little NEPA" requirements. They also permit a Federal agency to adopt another agency's EIS.

viii. *Consistent terminology.* The regulations provide a uniform terminology for the implementation of NEPA. For instance, the CEQ requirement for an environmental assessment will replace the following (nonexhaustive) list of comparable existing agency procedures: "survey" (Corps of Engineers), "environmental analysis" (Forest Service), "initial assessment" (Transportation), "normal or special clearance" (HUD), "environmental analysis report" (Interior), and "marginal impact statement" (HEW).)

ix. *Reducing paperwork require-*

ments. The regulations will reduce reporting paperwork requirements as summarized below. The existing Guidelines issued under Executive Order 11514 cover section 102(2)(C) of NEPA (environmental impact statements), and the new CEQ regulations cover sections 102(2) (A) through (I). The regulations replace not only the requirements of the Guidelines concerning environmental impact statements, but also replace more than 70 different sets of existing agency regulations, although each agency will issue its own implementing procedures to explain how these regulations apply to its particular programs.

Existing Requirements (Applicable Guidelines sections are noted.)	New Requirements (Applicable regulations sections are noted.)
Assessment (optional under Guidelines on a case-by-case basis; currently required, however by most major agencies in practice or in procedures) 1500.6.	*Assessment* (limited requirement: not required where there would not be environmental effects or where an EIS would normally be required) 1501.3, .4,
Notice of intent to prepare impact statement 1500.6.	*Notice of intent* to prepare EIS and commence scoping process 1501.7
Quarterly list of notices of intent 1500.6.	Requirement abolished.
Negative determination (decision not to prepare impact, statement) 1500.6.	*Finding of no significant impact* 1501.4.
Quarterly list of negative determinations 1500.6.	Requirement abolished.
Draft EIS 1500.7	*Draft EIS* 1502.9
Final EIS 1500.6, .10	*Final EIS* 1502.9
EISs on legislative reports ("agency reports on legislation initiated elsewhere") 1500.5(a)(1).	Requirement abolished.
Agency report to CEQ on implementation experience 1500.14(b).	Do.

Existing Requirements (Applicable Guidelines sections are noted.)	New Requirements (Applicable regulations sections are noted.)
Agency report to CEQ on substantive guidance 1500.6(c), .14.	Do.
Record of decision (no Guideline provision but required by many agencies' own procedures and in a wide range of cases generally under the Administrative Procedure Act and OMB Circular A-95, Part I, sec. 6(c) and (d), Part II, sec. 5(b)(4)).	*Record of decision* (brief explanation of decision EIS has been prepared; no circulation requirement) 1505.2.

B. REDUCING DELAY

The measures to reduce delay are

listed in § 1500.5 of the regulations.

i. *Time limits on the NEPA process.* The regulations encourage lead agencies to set time limits on the NEPA process and require that they be set when requested by an applicant.

ii. *Integrating EIS requirements with other environmental review requirements.* Often the NEPA process and the requirements of other laws proceed separately, causing delay. The regulations provide for all agencies with jurisdiction over the project to cooperate so that all reviews may be conducted simultaneously.

iii. *Integrating the NEPA process into early planning.* If environmental review is tacked on to the end of the planning process, then the process is prolonged, or else the EIS is written to justify a decision that has already been made, and genuine consideration may not be given to environmental factors.

iv. *Emphasizing interagency cooperation before the EIS is drafted.* The regulations emphasize that other agencies should begin cooperating with the lead agency before the EIS is prepared in order to encourage early resolution of differences. By having the affected agencies cooperate early in preparing a draft EIS, we hope both to produce a better draft and to reduce delays caused by unnecessarily late criticism.

v. *Swift and fair resolution of lead agency disputes.* When agencies differ as to who shall take the lead in preparing an EIS or none is willing to take the lead, the regulations provide a means for prompt resolution of the dispute.

vi. *Prepare EISs on programs and not repeat the same material in project specific EISs.* Material common to many actions may be covered in a broad EIS, and then through "tiering" may be incorporated by reference rather than reiterated in each subsequent EIS.

vii. *Legal delays.* The regulations provide that litigation should come at the end rather than in the middle of the process.

viii. *Accelerated procedures for legislative proposals.* The regulations provide accelerated simplified procedures for environmental analysis of legislative proposals, to fit better with Congressional schedules.

C. BETTER DECISIONS

Most of the features described above will help to improve decisionmaking. This, of course, is the fundamental purpose of the NEPA process, the end to which the EIS is a means. Section 101 of NEPA sets forth the substantive requirements of the Act, the policy to be implemented by the "action-forcing" procedures of Section 102. These procedures must be tied to their intended purpose, otherwise they are indeed useless paper work and wasted time. A central purpose of these regulations is to tie means to ends.

i. *Securing more accurate, professional documents.* The regulations insist upon accurate documents as the basis for sound decisions. The documents should draw upon all the appropriate disciplines from the natural and social sciences, plus the environmental design arts. The lead agency is responsible for the professional integrity of reports, and care should be taken to keep any possible bias from data prepared by applicants out of the environmental analysis. A list of people who helped prepare documents, and their professional qualifications, should be included in the EIS.

ii. *Recording in the decision how the EIS was used.* The new regulations require agencies to point out in the EIS analysis of alternatives which one is preferable on environmental grounds—including the often-overlooked alternative of no action at all. (However, if "no action" is identified as environmentally preferable, a second-best alternative must also be pointed out.)

Agencies must also produce a concise public record, indicating how the EIS was used in arriving at the decision. If the EIS is disregarded, it really is useless paperwork. It only contributes if it is used by the decisionmaker and the pubic. The record must state what the final decision was; whether the environmentally preferable alternative was selected; and if not, what considerations of national policy led to another choice.

iii. *Insure follow-up of agency decisions.* When an agency requires environmentally protective mitigation measures in its decision, the regulations provide for means to ensure that these measures are monitored and implemented.

Taken altogether, the regulations aim for a streamlined process, but one which as a broader purpose than the Guidelines they replace. The Guidelines emphasized a single document, the EIS, while the regulations emphasize the entire NEPA process, from early planning through assessment and EIS preparation through provisions for follow-up. They attempt to gear means to ends—to insure that the action-forcing procedures of sec. 102(2) of NEPA are used by agencies to fulfill the requirements of the Congressionally mandated policy set out in sec. 101 of the Act. Furthermore, the regulations are uniform, applying in the same way to all federal agencies, although each agency will develop its own procedures for implementing the regulations. Our attempt has been with these new regulations to carry out as faithfully as possible the original intent of Congress in enacting NEPA.

3. BACKGROUND

We have been greatly assisted in our task by the hundreds of people who responded to our call for suggestions on how to make the NEPA process work better. In public hearings which we held in June 1977, we invited testimony from a broad array of public officials, organizations, and private citizens, affirmatively involving NEPA's critics as well as its friends.

Among those represented were the U.S. Chamber of Commerce, which coordinated testimony from business; the Building and Construction Trades Department of the AFL–CIO, for labor; the National Conference of State Legislatures, for state and local governments; the Natural Resources Defense Council, for environmental groups. Scientists, scholars, and the general public were there.

There was extraordinary consensus among these diverse witnesses. All, without exception, expressed the view that NEPA benefited the public. Equally widely shared was the view that the process had become needlessly cumbersome and should be trimmed down. Witness after witness said that the length and detail of EIS's made it extremely difficult to distinguish the important from the trivial. The degree of unanimity about the good and bad points of the NEPA process was such that at one point an official spokesman for the oil industry rose to say that he adopted in its entirety the presentation of the President of the Sierra Club.

After the hearings we culled the record to organize both the problems and the solutions proposed by witnesses into a 38-page "NEPA Hearing Questionnaire." The questionnaire was sent to all witnesses, every state governor, all federal agencies, and everyone who responded to an invitation in the FEDERAL REGISTER. We received more than 300 replies, from a broad cross section of groups and individuals. By the comments we received from respondents we gauged our success in faithfully presenting the results of the public hearings. One commenter, an electric utility official, said that for the first time in his life he knew the government was listening to him, because all the suggestions made at the hearing turned up in the questionnaire. We then collated all the responses for use in drafting the regulations.

We also met with every agency of the federal government to discuss what should be in the regulations. Guided by these extensive interactions with government agencies and the public, we prepared draft regulations which were circulated for comment to all federal agencies in December 1977. We then studied agency comments in detail, and consulted numerous federal officials with special experience in implementing the Act. Informal redrafts were circulated to the agencies with greatest experience in preparing environmental impact statements. Improvements from our December 12 draft reflect this process.

At the same time that federal agencies were reviewing the early draft, we continued to meet with, listen to, and brief members of the public, including representatives of business, labor, state and local governments, environmental groups and others. We also considered seriously and proposed in our regulations virtually every major recommendation made by the Commission on Federal Paperwork and the General Accounting Office in their recent studies on the environmental impact statement process. The studies by these two independent bodies were among the most detailed and informed reviews of the paperwork abuses of the impact statement process. In many cases, such as streamlining intergovernmental coordination, the proposed regulations go further than their recommendations.

4. Exclusion

It should be noted that the issue of application of NEPA to environmental effects occurring outside the United States is the subject of continued discussions within the government and is not addressed in these regulations. Affected agencies continue to hold different views on this issue. Nothing in these regulations should be construed as asserting that NEPA either does or does not apply in this situation.

5. Analysis and Assessment of The Regulations

Since Executive Order 12044 became effective on March 23, 1978, after the Council's draft NEPA regulations had completed interagency review, the extent to which Executive Order 12044 applies to the Council's nearly completed process of developing NEPA regulations is not clear. Nevertheless, the requirements of Executive Order 12044 have been undertaken to the fullest extent possible. The analyses required by sections 2 (b), (c), (d), and 3(b), to the extent they may apply to the Council's proposed NEPA regulations, are available on request.

The Council has prepared a special environmental assessment of these regulations to illustrate the analysis that is appropriate under NEPA. The assessment discusses alternative regulatory approaches. Some regulations lend themselves to an analysis of their environmental impacts, particularly regulations with substantive requirements of those which apply to a physical setting. Although the Council obviously believes that its regulations will work to improve environmental quality, the impacts of procedural regulations of this kind are not susceptible to detailed analysis beyond that set out in the assessment.

Both the analyses under Executive Order 12044 and the assessment described above are available on request. Comments may be made on both documents in the same manner and by the same time as the comments on the regulations.

6. Additional Subjects for Comments

Several issues have been brought to our attention as appropriate subjects to be covered in the regulations. They are difficult issues on which we particularly solicit thoughtful views.

a. *Data bank.* Many were intrigued by the idea of a national data bank in which information developed in one EIS would be stored and become available for use in a subsequent EIS. Public comment on the questionnaire led us to conclude, reluctantly, that the idea is impractical. In practice most environmental information is specific to given areas or activities. To assemble a nationwide data bank would demand financial and other resources that are simply beyond the benefits that may be achieved. We have not included a data bank in these regulations but have instead tried to insure that in the scoping process the preparers of one EIS become aware of all related EISs so they can make use of the information in them. We would, however, welcome comment on this subject.

b. *Encouragement for agencies to fund public comments on EISs when an important viewpoint would otherwise not be presented.* The Council has been urged to provide either encouragement or direction to agencies, as part of their routine EIS preparation, to provide funds to responsible groups for public comments when important viewpoints would not otherwise be presented. Although we are acutely aware of the importance of comments to the success of the EIS process, we have not included such a provision. We would welcome comment on this subject also.

Conclusion

We look forward to your comments and help. To repeat, comments should be sent by August 11, 1978, to Nicholas C. Yost, General Counsel, Attention: NEPA Comments, Council on Environmental Quality, 722 Jackson Place NW., Washington, D.C. 20006.

Thank you for cooperating with us.

Charles Warren,
Chairman.

Title 40 Chapter V is proposed to be amended by revising Part 1500 and by adding Parts 1501 through 1508 to read as follows:

PART 1500—PURPOSE, POLICY, AND MANDATE

Sec.
1500.1 Purpose.
1500.2 Policy.
1500.3 Mandate.

1500.4 Reducing paperwork.
1500.5 Reducing delay.
1500.6 Agency authority.

AUTHORITY: NEPA, the Environmental Quality Improvement Act of 1970, as amended (42 U.S.C. 4371 et seq.), section 309 of the Clean Air Act, as amended (42 U.S.C. 1857h-7), and Executive Order 11514, Protection and Enhancement of Environmental Quality (March 5, 1970 as amended by Executive Order 11991, May 24, 1977).

§ 1500.1 Purpose.

(a) The National Environmental Policy Act (NEPA) is our basic national charter for protection of the environment. It establishes policy, sets goals (section 101), and provides means (section 102) for carrying out the policy. Section 102(2) contains "action-forcing" provisions to make sure that federal agencies act according to the letter and spirit of the Act. The regulations that follow implement Section 102(2). Their purpose is to tell federal agencies what they must do to comply with the procedures and achieve the goals of the Act. The President, the federal agencies, and the courts share responsibility for enforcing the Act so as to achieve the substantive requirements of section 101.

(b) NEPA procedures must insure that environmental information is available to public officials and citizens before decisions are made and before actions are taken. The information must be of high quality. Accurate scientific analysis, expert agency comments, and public scrutiny are essential to implementing NEPA. Most important, NEPA documents must concentrate on the issues that are truly significant to the action in question, rather than amassing needless detail.

(c) Ultimately, of course, it is not better documents but better decisions that count. NEPA's purpose is not generate paperwork—even excellent paperwork—but to foster excellent action. The NEPA process is intended to help public officials make decisions that are based on understanding of environmental consequences, and take actions that protect, restore, and enhance the environment. These regulations provide the direction to achieve this purpose.

§ 1500.2 Policy.

Federal agencies shall to the fullest extent possible:

(a) Interpret and administer the policies, regulations, and public laws of the United States in accordance with the policies set forth in the Act and in these regulations.

(b) Implement procedures to make the NEPA process more useful to decisionmakers and the public; to reduce paperwork and the accumulation of extraneous background data; and to emphasize real environmental issues and alternatives. Environmental impact statements shall be concise, clear, and to the point, and shall be supported by evidence that agencies have made the necessary environmental analyses.

(c) Integrate the requirements of NEPA with other planning and environmental review procedures required by law or by agency practice so that all such procedures run concurrently, rather than consecutively.

(d) Encourage and facilitate public involvement in decisions which affect the quality of the human environment.

(e) Use the NEPA process to identify and assess the reasonable alternatives to proposed actions that will avoid or minimize adverse effects of these actions upon the quality of the human environment.

(f) Use all practicable means, consistent with the requirements of the Act and other essential considerations of national policy, to restore and enhance the quality of the human environment and avoid or minimize any possible adverse effects of their actions upon the quality of the human environment.

§ 1500.3 Mandate.

Parts 1500-1508 of this Title provide regulations applicable to and binding on all Federal agencies for implementing the procedural provisions of the National Environmental Policy Act of 1969, as amended (Pub. L. 91-190, 42 U.S.C. 4321 et seq.) (NEPA or the Act) except where compliance would be inconsistent with other statutory requirements. These regulations are issued pursuant to NEPA, the Environmental Quality Improvement Act of 1970, as amended (42 U.S.C. 4371 et seq.), Section 309 of the Clean Air Act, as amended (42 U.S.C. 1857h-7), and Executive Order 11514, Protection and Enhancement of Environmental Quality (March 5, 1970, as amended by Executive Order 11991, May 24, 1977). These regulations, unlike the predecessor guidelines, are not confined to Sec. 102(2)(C) (environmental impact statements). The regulations apply to the whole of section 102(2). The provisions of the Act and of these regulations must be read together as a whole in order to comply with the spirit and letter of the law. It is the Council's intention that judicial review of agency compliance with these regulations not occur before an agency has filed the final environmental impact statement, or has made a finding of no significant impact, or takes action that will result in irreparable injury.

§ 1500.4 Reducing paperwork.

Agencies shall reduce excess paperwork by:

(a) Reducing the length of environmental impact statements (§ 1502.2(c)), by means such as setting appropriate page limits (§ 1501.7(b)(1) and 1502.7).

(b) Preparing analytic rather than encyclopedic environmental impact statements (§ 1502.2(a)).

(c) Discussing only briefly issues other than significant ones (§ 1502.2(b)).

(d) Writing environmental impact statements in plain language (§ 1502.8).

(e) Following a clear format for environmental impact statements (§ 1502.10).

(f) Emphasizing the portions of the environmental impact statement that are useful to decisionmakers and the public (§§ 1502.14 and 1502.15) and reducing emphasis on background material (§ 1502.16).

(g) Using the scoping process not only to identify significant environmental issues deserving of study, but also to deemphasize insignificant issues, narrowing the scope of the environmental impact statement process accordingly (§ 1501.7).

(h) Summarizing the environmental impact statement (§ 1502.12) and circulating the summary instead of the entire environmental impact statement if the latter is unusually long (§ 1502.19).

(i) Using program, policy, or plan environmental impact statements and tiering from statements of broad scope to those of narrower scope to eliminate repetitive discussions of the same issues (§§ 1502.4 and 1502.20).

(j) Incorporating by reference (§ 1502.21).

(k) Integrating NEPA requirements with other environmental review and consultation requirements (§ 1502.25).

(l) Requiring comments to be as specific as possible (§ 1503.3).

(m) Attaching and circulating only changes to the draft environmental impact statement, rather than rewriting and circulating the entire statement when changes are minor (§ 1503.4(b)).

(n) Eliminating duplication with State and local procedures by providing for joint preparation (§ 1506.2) and with other Federal procedures by providing for one agency's adoption of appropriate environmental documents prepared by another agency (§ 1506.3).

(o) Combining environmental documents with other documents (§ 1506.4).

(p) Using categorical exclusions to exclude from environmental impact statement requirements categories of actions which do not individually or cumulatively have a significant effect on the human environment (§ 1508.4).

(q) Using a finding of no significant impact and not preparing an environmental impact statement when an action not otherwise excluded will not have a significant effect on the human environment (§ 1508.13).

§ 1500.5 Reducing delay.

Agencies shall reduce delay by:

(a) Integrating the NEPA process into early planning (§ 1501.2).

(b) Emphasizing interagency cooperation before the environmental impact statement is prepared rather than adversary comments on a completed document (§ 1501.6).

(c) Insuring the swift and fair resolution of lead agency disputes (§ 1501.5).

(d) Using the scoping process for an early identification of what are and what are not the real issues (§ 1501.7).

(e) Establishing appropriate time limits for the environmental impact statement process (§§ 1501.7(b)(2) and 1501.8).

(f) Preparing environmental impact statements early in the process (§ 1502.5).

(g) Integrating NEPA requirements with other environmental review and consultation requirements §§ 1502.25).

(h) Eliminating duplication with State and local procedures by providing for joint preparation (§ 1506.2) and with other Federal procedures by providing for one agency's adoption of appropriate environmental documents prepared by another agency (§ 1506.3).

(i) Combining environmental documents with other documents (§ 1506.4).

(j) Using accelerated procedures for proposals for legislation (§ 1506.8).

(k) Using categorical exclusions to exclude from environmental impact statement requirements categories of actions which do not individually or cumulatively have a significant effect on the human environment (§ 1508.4).

(l) Using a finding of no significant impact and not preparing an environmental impact statement when an action not otherwise excluded will not have a significant effect on the human environment (§ 1508.13).

§ 1500.6 Agency authority.

Each agency shall interpret the provisions of the Act as a supplement to its existing authority and as a mandate to view traditional policies and missions in the light of the Act's national environmental objectives. Agencies shall review their policies, procedures, and regulations accordingly and revise them as necessary to ensure full compliance with the purposes and provisions of the Act. The phrase "to the fullest extent possible" in section 102 means that each agency of the Federal Government shall comply with that section unless existing law applicable to the agency's operations expressly prohibits or makes compliance impossible.

PART 1501—NEPA AND AGENCY PLANNING

AUTHORITY: NEPA, the Environmental Quality Improvement Act of 1970, as amended (42 U.S.C. 4371 et seq.), Section 309 of the Clean Air Act, as amended (42 U.S.C. 1857h-7), and Executive Order 11514, Protection and Enhancement of Environmental Quality (March 5, 1970, as amended by Executive Order 11991, May, 24, 1977).

§ 1501.1 Purpose.

The purposes of this part include:

(a) Integrating the NEPA process into early planning to insure appropriate consideration of NEPA's policies and to eliminate delay.

(b) Emphasizing cooperative consultation among agencies before the environmental impact statement is prepared rather than adversary comments on a completed document.

(c) Providing for the swift and fair resolution of lead agency disputes.

(d) Identifying at an early stage the significant environmental issues deserving of study and deemphasizing insignificant issues, narrowing the scope of the environmental impact statement accordingly.

(e) Providing a mechanism for putting appropriate time limits on the environmental impact statement process.

§ 1501.2 Apply NEPA early in process.

Agencies shall integrate the NEPA process with other planning at the earliest possible time to insure that planning and decisions reflect environmental values, to avoid delays later in the process, and to head off potential conflicts. Each agency shall:

(a) As specified by § 1507.2 comply with the mandate of sec. 102(2)(A) to "utilize a systematic, interdisciplinary approach which will insure the integrated use of the natural and social sciences and the environmental design arts in planning and in decisionmaking which may have an impact on man's environment."

(b) Identify environmental effects and values in adequate detail so they can be compared to economic and technical analyses. Environmental documents and appropriate analyses shall be circulated and reviewed at the same time as other planning documents.

(c) Study, develop, and describe appropriate alternatives to recommended courses of action in any proposal which involves unresolved conflicts concerning alternative uses of available resources as provided by sec. 102(2)(E) of the Act.

(d) Provide for cases where actions are planned by other than Federal agencies before Federal involvement so that:

(1) The sponsor of the proposal initiates studies if Federal involvement is foreseeable.

(2) The Federal agency consults early with appropriate State and local agencies and with interested private persons and organizations when its own involvement is reasonably foreseeable.

(3) The Federal agency commences its NEPA process at the earliest possible time.

§ 1501.3 When to prepare an environmental assessment.

An environmental assessment (§ 1508.9) shall be prepared unless one is not necessary under the procedures adopted under § 1507.3(b). Agencies may prepare an assessment on any action at any time in order to assist agency planning and decisionmaking.

§ 1501.4 Whether to prepare an environmental impact statement.

In determining whether to prepare an environmental impact statement the Federal agency shall:

(a) Determine under § 1507.3 whether the proposal is one which

(1) Normally requires an environmental impact statement, or

(2) Normally does not require either an environmental impact statement or an environmental assessment (categorical exclusion).

(b) If the proposed action is not covered by paragraph (a), prepare an environmental assessment (§ 1508.9). The agency shall involve environmental agencies and the public, to the extent practicable, in preparing the assessment.

(c) Based on the environmental assessment make its determination whether to prepare an environmental impact statement.

(d) If the agency will prepare an environmental impact statement, the agency shall commence the scoping process (§ 1501.7).

(e) If the agency determines on the basis of the environmental assessment not to prepare a statement, the agency shall prepare a finding of no significant impact (§ 1508.13).

(1) The agency shall make the finding of no significant impact available in a manner calculated to inform the affected public as specified in § 1506.6.

(2) In certain limited circumstances the agency shall make the finding of no significant impact available for public review for 30 days before the agency makes its final determination

whether to prepare an environmental impact statement and before the action may begin. The circumstances are:

(i) The proposed action is, or is closely similar to, one which normally requires the preparation of an environmental impact statement under the procedures adopted by the agency pursuant to section 1507.3(b), or

(ii) The nature of the proposed action is one without precedent.

§ 1501.5 Lead agencies.

(a) A lead agency shall supervise the preparation of an environmental impact statement if more than one Federal agency either:

(1) Proposes or is involved in the same action; or

(2) Is involved in a group of actions directly related to each other because of their functional interdependence or geographical proximity.

(b) More than one Federal, State, or local agency, one of which must be Federal, may act as joint lead agencies to prepare an environmental impact statement (section 1506.2).

(c) If an action satisfies the provisions of paragraph (a) of this section the potential lead agencies concerned shall determine by letter or memorandum which agency shall be the lead agency and which shall be cooperating agencies. The agencies shall resolve the lead agency question in a manner that will not cause delay. If there is disagreement among the agencies, the following factors (which are listed in descending importance) shall determine lead agency designation:

(1) Magnitude of agency's involvement.

(2) Project approval/disapproval authority.

(3) Expertise concerning the action's environmental effects.

(4) Duration of agency's involvement.

(5) Sequence of agency's involvement.

(d) If potential lead agencies fail to agree on which agency shall be the lead agency as specified in paragraph (c) of this section, (1) any Federal agency or (2) any State or local agency or private person substantially affected by the absence of agreement on lead agency designation may make a written request to the potential lead agencies that a lead agency be designated.

(e) If Federal agencies are unable to agree on which agency will be the lead agency or if the procedure described in paragraph (d) of this section has not resulted within a reasonable time in a lead agency designation, any of the agencies or persons concerned may file a request with the Council asking it to determine which Federal agency shall be the lead agency.

A copy of the request shall be transmitted to each potential lead agency. The request shall consist of:

(1) A precise description of the nature and extent of the proposed action;

(2) A detailed statement of why each potential lead agency should or should not be the lead agency under the criteria specified in subparagraph (2).

(f) A response may be filed by any potential lead agency concerned within 20 days after a request is filed with the Council. The Council shall determine within 20 days after receiving the request and all responses which Federal agency shall be the lead agency and the extent to which the other Federal agencies concerned shall be cooperating Federal agencies.

§ 1501.6 **Cooperating agencies.**

The purpose of this section is to emphasize agency cooperation early in the NEPA process. Upon request of the lead agency, any other Federal agency which has jurisdiction by law shall be a cooperating agency. In addition any other Federal agency which has special expertise with respect to any environmental issue, which should be addressed in the statement may be a cooperating agency upon request of the lead agency.

(a) The lead agency shall:

(1) Request the participation of each cooperating agency in the NEPA process at the earliest possible time.

(2) To the maximum extent possible consistent with its responsibility as lead agency use the environmental analysis and proposals of cooperating agencies with jurisdiction by law or special expertise.

(3) Meet with a cooperating agency at the latter's request.

(b) Each cooperating agency shall:

(1) Participate in the NEPA process at the earliest possible time.

(2) Participate in the scoping process.

(3) Assume on request of the lead agency responsibility for developing information and preparing environmental analyses including portions of the environmental impact statement concerning which the cooperating agency has special expertise.

(4) Make available staff support at the lead agency's request to enhance the latter's interdisciplinary capability.

(5) Normally a cooperating agency shall use its own funds. The lead agency shall, to the extent available funds permit, fund those major activities or analyses it requests from cooperating agencies. Potential lead agencies shall include such funding requirements in their budget requests

§ 1501.7 **Scoping.**

There shall be an early and open process for determining the scope of issues to be addressed and for identifying the significant issues. This process shall be termed scoping. As soon as practicable after its decision to prepare an environmental impact statement and before the scoping process the lead agency shall publish a notice of intent (§ 1508.21) in the FEDERAL REGISTER.

(a) As part of the scoping process the lead agency shall:

(1) Invite the participation of affected Federal, State, and local agencies, any affected Indian tribe, the proponent of the action, and other interested persons (including those who might not be in accord with the action).

(2) Determine the scope (§ 1508.24) and the significant issues to be analyzed in depth in the environmental impact statement.

(3) Identify and eliminate from detailed study the issues which are not significant or which have been covered by prior environmental review (§ 1506.3), narrowing the discussion of these issues in the statement to a brief presentation of why they will not have a significant effect on the human environment or a reference to their coverage elsewhere.

(4) Allocate assignments for preparation of the environmental impact statement among the lead and cooperating agencies, with the lead agency retaining responsibility for the statement.

(5) Indicate any environmental assessments and other environmental impact statements which are being or

will be prepared that are related to but are not part of the scope of the impact statement which is the subject of the meeting.

(6) Identify other environmental review and consultation requirements so the lead and cooperating agencies may comply with section 1502.25.

(7) Indicate the relationship between the timing of the preparation of environmental analyses and the agency's tentative planning and decision-making schedule.

(8) When practicable hold an early scoping meeting or meetings which may be integrated with any other early planning meeting the agency has. Such scoping meeting will often be appropriate when the impacts of a particular action are confined to specific sites.

(b) As part of the scoping process the lead agency may:

(1) Set page limits on environmental documents (§ 1502.7).

(2) Set time limits (§ 1501.8).

(c) An agency shall revise the determinations made under paragraphs (a) and (b) of this section if substantial changes are made later in the proposed action or if significant new circumstances (including information) arise which bear on the proposal or its impacts.

§ 1501.8 Time limits.

Although the Council has decided that universal time limits for the entire NEPA process are too inflexible to prescribe, Federal agencies are encouraged to set time limits appropriate to individual action (consistent with § 1506.10). When multiple agencies are involved the reference to agency below means lead agency.

(a) The agency shall:

(1) Consider the following factors in determining time limits:

(i) Potential for environmental harm.

(ii) Size of the proposed action.

(iii) State of the art of analytic techniques.

(iv) Degree of public need for the proposed actions, including the consequences of delay.

(v) Number of persons and agencies affected.

(vi) Degree to which relevant information is known and if not known the time required for obtaining it.

(vii) Degree to which the action is controversial.

(2) Set limits if an applicant for the proposed action requests them, provided that they are consistent with the purposes of NEPA and other essential considerations of national policy.

(b) The agency may:

(1) Set overall time limits or limits for each constituent part of the NEPA process, which may include:

(i) Decision on whether to prepare an environmental impact statement (if not already decided).

(ii) Determination of the scope of the environmental impact statement.

(iii) Preparation of the draft environmental impact statement.

(iv) Review of any comments on the draft environmental impact statement from the public and agencies.

(v) Preparation of the final environmental impact statement.

(vi) Review of any comments on the final environmental impact statement.

(vii) Decision on the action based in part on the environmental impact statement.

(2) Designate a person (such as the project manager or a person in the agency's office with NEPA responsibilities) to expedite the NEPA process.

(c) State or local agencies or members of the public may request a Federal Agency to set time limits.

PART 1502—ENVIRONMENTAL IMPACT STATEMENT

1502.21 Incorporation by Reference.
1502.22 Incomplete or Unavailable Information.
1502.23 Cost-Benefit Analysis.
1502.24 Methodology and Scientific Accuracy.
1502.25 Environmental Review and Consultation Requirements.

AUTHORITY: NEPA, the Environmental Quality Improvement Act of 1970, as amended (42 U.S.C. 4371 et seq.), Section 309 of the Clean Air Act, as amended (42 U.S.C. 1857h-7), and Executive Order 11514, Protection and Enhancement of Environmental Quality (March 5, 1970, as amended by Executive Order 11991, May 24, 1977).

§ 1502.1 Purpose.

The primary purpose of an environmental impact statement is as an action-forcing device to insure that the policies and goals defined in the Act are infused into the ongoing programs and actions of the Federal Government. It shall provide full and fair discussion of significant environmental impacts and shall inform decisionmakers and the public of the reasonable alternatives which would avoid or minimize adverse impacts or enhance the quality of the human environment. Agencies shall focus on significant environmental issues and alternatives and shall reduce paperwork and the accumulation of extraneous background data. Statements shall be concise, clear, and to the point, and shall be supported by evidence that the agency has made the necessary environmental analyses. An environmental impact statement is more than a disclosure document. It shall be used by Federal officials in conjunction with other relevant material to plan actions and make decisions.

§ 1502.2 Implementation.

To achieve the purposes set forth in § 1502.1 agencies shall prepare environmental impact statements in the following manner:

(a) Environmental impact statements shall be analytic rather than encyclopedic.

(b) Impacts shall be discussed in proportion to their significance. There shall be only brief discussion of other than significant issues. As in a finding of no significant impact, there should be only enough discussion to show why more study is not warranted.

(c) Environmental impact statements shall be kept concise and shall be no longer than absolutely necessary to comply with NEPA with these regulations. Length should vary first with potential environmental problems and then with project size.

(d) Environmental impact statements shall state how alternatives considered in it and decisions baseon on it will or will not achieve the requirements of sections 101 and 102(1) of the Act and other environmental laws and policies.

(e) The range of alternatives discussed in environmental impact statements shall encompass those the ultimate agency decisionmaker considers.

(f) Agencies shall not commit resources prejudicing selection of alternatives before making a final decision (§ 1506.1).

(g) Environmental impact statements shall serve as the means of assessing the environmental impact of proposed agency actions, rather than justifying decisions already made.

§ 1502.3 Statutory requirements for statements.

As required by sec. 102(2)(C) of NEPA environmental impact statements (§ 1508.11) are to be included in every recommendation or report
On proposals (§ 1508.22).
For legislation and (§ 1508.16).
Other major Federal actions (§ 1508.17).
Significantly (§ 1508.25).
Affecting (§§ 1508.3, 1508.8).
The quality of the human environment (§ 1508.14).

§ 1502.4 Major Federal actions requiring the preparation of environmental impact statements.

(a) Agencies shall make sure the proposal which is the subject of an environmental impact statement is properly defined. Agencies shall use the criteria for scope (§ 1508.24) to determine which proposal(s) shall be the subject of a particular statement. Proposals or parts of proposals which are related to each other closely enough to be, in effect, a single course of action shall be evaluated in a single impact statement.

(b) Environmental impact statements may be prepared, and are sometimes required, for broad Federal actions such as the adoption of new agency programs or regulations (§ 1508.17). Agencies shall prepare statements on broad actions to be relevant to policy and timed to coincide with meaningful points in agency planning and decisionmaking.

(c) When preparing statements on broad actions, agencies may find it useful to evaluate the proposal(s) by one or more agencies in one of the following ways:

(1) Geographic, including actions occurring in the same general location, such as an ocean, region, or metropolitan area.

(2) Generic, including actions which have relevant similarities, such as common timing, impacts, alternatives, methods of implementation, media, or subject matter.

(3) Technological development including federal or federally assisted research, development or demonstration programs aimed at developing new technologies which, if applied, could significantly affect the quality of the human environment. Statements shall be prepared on such programs and shall be available before the program has reached a stage of investment or commitment to implementation likely to determine subsequent development or restrict later alternatives.

(d) Agencies shall as appropriate employ scoping (§ 1501.7), tiering (§ 1502.20), and other methods listed in §§ 1500.4 and 1500.5 to relate broad and narrow actions and to avoid duplication and delay.

§ 1502.5 Timing.

An agency shall commence preparation of an environmental impact statement as close as possible to the time the agency makes or is presented with a proposal (§ 1508.22) so that preparation can be completed in time for the final statement to be included in any recommendation or report on the proposal. The statement shall be prepared early enough so that it can practically serve as an important contribution to the decisionmaking process and shall not be used to rationalize or justify decisions already made (§§ 1500.2(c), 1501.2, and 1502.2). For instance:

(a) For projects directly undertaken by Federal agencies such statements shall be prepared at the feasibility analysis (go-no go) stage and may be supplemented at a later stage if necessary

(b) For applications to the agency appropriate preliminary environmental assessments or statements shall be commenced at the latest immediately after the application is received, but federal agencies are encouraged to prepare them earlier, preferably jointly with applicable State or local agencies.

(c) For adjudication, the final environmental impact statement shall normally precede the final staff recommendation and that portion of the public hearing related to the impact study. In appropriate circumstances the statement may follow preliminary hearings designed to gather information for use in the statements.

§ 1502.6 Interdisciplinary preparation.

Environmental impact statements shall be prepared using an inter-disciplinary approach which will insure the integrated use of the natural and social sciences and the environmental design arts (section 102(2)(A) of the Act). The disciplines of the preparers shall be correlated to the scope and issues identified in the scoping process (§ 1501.7).

§ 1502.7 Page limits.

The text of final environmental impact statements (e.g., paragraphs (d) through (g) of § 1502.10) shall normally be less than 150 pages and for proposals of unusual scope or complexity shall normally be less than 300 pages.

§ 1502.8 Writing.

Environmental impact statements shall be written in plain language and may use appropriate graphics so that they may be understood by decisionmakers and the public. Agencies should employ writers of clear prose or editors to write, review, or edit statements, which will be based upon the analysis and supporting data from the natural and social sciences and the environmental design arts.

§ 1502.9 Draft, final, and supplemental statements.

Except as provided in § 1506.8, environmental impact statements shall be prepared in two stages and may be supplemented.

(a) Draft environmental impact statements shall be prepared in accordance with the scope decided upon in the scoping process. The lead agency shall work with the cooperating agencies and shall obtain comments as required in Part 1503. At the time the draft statement is prepared it must fulfill and satisfy to the fullest extent possible the requirements established for final statements in section 102(2)(C) of the Act. If a draft statement is so inadequate as to preclude meaningful analysis, the agency shall prepare and circulate a revised draft of the appropriate portion. In the draft statement the agency shall make every effort to disclose and discuss at appropriate points in the text all major points of view on the environmental impacts of the alternatives including the proposed action.

(b) Final environmental impact statements shall respond to comments as required in Part 1503. In the final statement the agency shall discuss at appropriate points in the text the existence of any responsible opposing view not adequately discussed in the draft statement and shall indicate the agency's response to the issues raised.

(c) Agencies:

(1) Shall prepare supplements to either draft or final environmental impact statements if:

(i) The agency makes substantial changes in the proposed action that are relevant to environmental concerns; or

(ii) There are significant new circumstances, relevant to environmental concerns (including information), bearing on the proposed action or its impacts.

(2) May also prepare supplements when the agency determines that the purposes of the Act will be furthered by doing so.

(3) Shall adopt procedures for introducing a supplement into its formal administrative record, if such a record exists.

(4) Shall prepare, circulate, and file a supplement to a statement in the same fashion (exclusive of scoping) as a draft statement unless alternative procedures are approved by the Council.

§ 1502.10 Recommended format.

Agencies shall use a format for environmental impact statements which will encourage good analysis and clear presentation of the alternatives including the proposed actions. The following standard format for environmental impact statements should be followed unless there is a compelling reason to do otherwise:

(a) Cover sheet

(b) Summary

(c) Table of Contents

(d) Purpose of and Need for Action

(e) Alternatives Including Proposed Action (secs. 102(2)(C)(iii) and 102(2)(E) of the Act).

(f) Environmental Consequences (especially secs. 102(2)(C) (i), (ii), (iv), and (v) of the Act.

(g) Affected Environment.

(h) List of Preparers.

(i) List of Agencies, Organizations, and Persons to Whom Copies of the Statement Are Sent.

(j) Index.

(k) Appendices (if any).

If a different format is used, it shall include paragraphs (a), (b), (c), (h), (i), and (j), of this section and shall include the substance of paragraphs (d), (e), (f), (g), and (k) of this section as further described in §§ 1502.11–1502.18 in any appropriate format.

§ 1502.11 Cover sheet.

The cover sheet shall not exceed one page. It shall include:

(a) The name of the responsible agencies including the lead agency and any cooperating agencies.

(b) The name of the proposed action that is the subject of the statement (and if appropriate the names of related cooperating agency actions), together with the State(s) and county(ies) (or the country if applicable) where the action is located.

(c) The name, address, and telephone number of the person at the agency who can supply further information.

(d) A designation of the statement as a draft, final, or draft or final supplement.

(e) A one paragraph abstract of the statement.

(f) The date by which comments must be received (computed in cooperation with EPA § 1506.10)).

§ 1502.12 Summary.

Each environmental impact statement shall contain a summary which adequately and accurately summarizes the statement. The summary shall stress the major conclusions, areas of controversy (including issues raised by agencies and the public), and the issues to be resolved (including the choice among alternatives). The summary will normally not exceed 15 pages.

§ 1502.13 **Purpose and need.**

The statement shall briefly specify the underlying purpose and need to which the agency is responding in proposing the action and alternatives. Normally this section shall not exceed one page.

§ 1502.14 **Alternatives including the proposed action.**

This section is the heart of the environmental impact statement. Based on the information and analysis presented in the sections on the Environmental Consequences (§ 1502.15) and the Affected Environment (§ 1502.16), it should present the environmental impacts of the proposal and the alternatives in comparative form, thus sharpening the issues and providing a clear basis for choice among options by the decisionmaker and the public. In this section agencies shall:

(a) Rigorously explore and objectively evaluate all reasonable alternatives, and for alternatives which were eliminated from detailed study, briefly discuss the reasons for such elimination.

(b) Devote substantial treatment to each alternative considered in detail including the proposed action so that reviewers may evaluate the comparative merits.

(c) Include reasonable alternatives not within the jurisdiction of the lead agency.

(d) Include the no action alternative.

(e) Identify the environmentally preferable alternative (or alternatives if two or more are equally preferable) and the reasons for identifying it. If the alternative identified is for no action, the agency shall also identify the alternative other than no action that is environmentally preferable and the reasons for identifying it.

(f) Identify the agency's preferred alternative or alternatives if one or more exists in the draft statement and identify such alternative(s) in the final statement unless another law prohibits the expression of such a preference.

(g) Include appropriate mitigation measures not already included in the proposed action or alternatives.

§ 1502.15 **Environmental consequences.**

This section forms the scientific and analytic basis for the comparisons under § 1502.14. It shall consolidate the discussions of those elements required by secs. 102(2)(C) (i), (ii), (iv), and (v) of NEPA which are within the scope of the statement and as much of sec. 102(2)(C)(iii) as is necessary to support the comparisons. This includes the environmental impacts of the proposed action and alternatives, any adverse environmental effects which cannot be avoided should the proposal be implemented, the relationship between short-term uses of man's environment and the maintenance and enhancement of long-term productivity, and any irreversible or irretrievable commitments of resources which would be involved in the proposed action should it be implemented. The Council intends that preparers not cause duplication in the discussions under § 1502.14 and this section. This section shall include discussions of:

(a) Direct effects and their significance (§1508.8).

(b) Indirect effects and their significance (§ 1508.8).

(c) Possible conflicts between the proposed action and the objectives of Federal, regional, State, and local land use plans, policies, and controls for the area concerned.

(d) The environmental effects of alternatives including the proposed action. The comparisons under § 1502.14 will be based on this discussion.

(e) Energy requirements and conservation potential of various alternatives and mitigation measures.

(f) Natural or depletable resource requirements and conservation potential of various alternatives and mitigation measures.

(g) Means to mitigate adverse environmental impacts (if not fully covered under § 1502.14(g)).

§ 1502.16 **Affected environment.**

The environmental impact statement shall succinctly describe the environment of the area or areas to be affected by the alternatives under consideration. The descriptions shall be no longer than is necessary to understand the effects of the alternatives. Data and analyses in a statement shall be commensurate with the importance of the impact, with less important material summarized, consolidated, or simply referenced. Agencies shall avoid useless bulk in statements and shall concentrate effort and attention on important issues. Verbose descriptions of the affected environment are themselves no measure of the adequacy of an environmental impact statement.

§ 1502.17 List of preparers.

The environmental impact statement shall list the names, together with their qualifications and professional disciplines (§ 1502.6 and 1502.8), of the persons who were primarily responsible for preparing the environmental impact statement or significant background papers, including basic components of the statement. Where possible the names of persons who are responsible for a particular analysis, including analyses in background papers, shall be identified. Normally the list will not exceed two pages.

§ 1502.18 Appendix.

If an agency prepares an appendix to an environmental impact statment the appendix shall:

(a) Consist of material prepared in connection with an environmental impact statement (as distinct from material which is not so prepared and which is incorporated by reference §1502.21)).

(b) Normally consist of material which substantiates any analysis fundamental to the impact statement.

(c) Normally be analytic and relevant to the decision to be made.

(d) Be circulated with the environmental impact statement or be readily available on request.

§ 1502.19 Circulation of the environmental impact statement.

Agencies shall circulate the entire draft and final environmental impact statements except as provided in § 1502.18(d) and 1503.4(c). However, if the statement is unusually long, the agency may circulate the summary instead, except that the entire statement shall be furnished to:

(a) Any Federal agency which has jurisdiction by law or special expertise with respect to any environmental impact involved and any appropriate Federal, State or local agency authorized to develop and enforce environmental standards.

(b) Any person, organization, or agency requesting the entire environmental impact statement.

(c) In the case of a final environmental impact statement any person, organization, or agency which submitted substantive comments on the draft.

If the agency circulates the summary and thereafter receives a timely request for the entire statement, the time for comment for that requestor only shall be extended by at least 15 days beyond the minimum period.

§ 1502.20 Tiering.

Agencies are encouraged to tier their environmental impact statements to eliminate repetitive discussions of the same issues and to focus on the actual issues ripe for decision at each level of environmental review (§ 1508.26). Whenever a broad environmental impact statement has been prepared (such as a program or policy statement) and a subsequent statement or environmental assessment is then prepared on an action included within the entire program or policy (such as a site specific action) the subsequent statement or environmental assessment need only summarize the issues discussed in the broader statement and incorporate such discussions by reference and shall concentrate on the issues specific to the subsequent action. Tiering may also be appropriate for different stages of actions. (Section 1508.26).)

§ 1502.21 Incorporation by reference.

Agencies shall incorporate material into an environmental impact statement by reference when to do so will cut down on bulk without impeding agency and public review of the action. The incorporated material shall be cited in the statement and its content briefly described. No material may be incorporated by reference unless it is reasonably available for inspection by potentially interested persons within the time allowed for comment. Material based on proprietary data which is itself not available for review and comment shall not be incorporated by reference.

§ 1502.22 Incomplete or unavailable information.

Agencies dealing with gaps in relevant information including scientific uncertainty, shall always make clear that such information is lacking or that uncertainty exists.

(a) If the information is essential to

a reasoned choice among alternatives and is not known and the costs of obtaining it are not exorbitant, the agency shall include the information in the environmental impact statement.

(b) If the information is important to the decision and the means to obtain it are not known (e.g., the means for obtaining it are beyond the state of the art) the agency shall weigh the need for the action against the risk and severity of possible adverse impacts were the action to proceed in the face of uncertainty. If the agency proceeds, it shall include a worst case analysis.

§ 1502.23   Cost-benefit analysis.

If a cost-benefit analysis is being considered for the proposed action, it shall be incorporated by reference or appended to the statement as an aid in evaluating the environmental consequences. To assess the adequacy of compliance with sec. 102(2)(B) of the Act the statement shall when a cost-benefit analysis is prepared discuss the relationship between that analysis and any analyses of unquantified environmental impacts, values, and amenities.

§ 1502.24   Methodology and scientific accuracy.

Agencies shall insure the professional, including scientific, integrity of the discussions and analyses in environmental impact statements. They shall identify any methodologies used and shall make explicit reference by footnote to the scientific and other sources relied upon for conclusions in the statement.

§ 1502.25   Environmental review and consultation requirements.

To the fullest extent possible, agencies shall prepare draft environmental impact statements concurrently with and integrated with environmental impact analyses and related surveys and studies required by the Fish and Wildlife Coordination Act (16 U.S.C. Sec. 661 et seq.) the National Historic Preservation Act of 1966 (16 U.S.C. Sec. 470 et seq.), the Endangered Species Act of 1972 (16 U.S.C. Sec. 1531 et seq.) and other environmental review laws.

PART 1503—COMMENTING

Sec.
1503.1   Inviting Comments.
1503.2   Duty to Comment.
1503.3   Specificity of Comments.
1503.4   Response to Comments.

AUTHORITY: NEPA, the Environmental Quality Improvement Act of 1970, as amended (42 U.S.C. 4371 et seq.), Section 309 of the Clean Air Act, as amended (42 U.S.C. 1857h-7), and Executive Order 11514, Protection and Enhancement of Environmental Quality (March 5, 1970, as amended by Executive Order 11991, May 24, 1977).

§ 1503.1   Inviting comments.

(a) After preparing a draft environmental impact statement and before preparing a final environmental impact statement the agency shall:

(1) Obtain the comments of any Federal agency which has jurisdiction by law or special expertise with respect to any environmental impact involved or which is authorized to develop and enforce environmental standards.

(2) Request the comments of appropriate State and local agencies which are authorized to develop and enforce environmental standards, or any agency which has requested that it receive statements on actions of the kind proposed.

(3) Request comments from the public, affirmatively soliciting comments from those persons or organizations who may be interested or affected.

(b) After preparing a final environmental impact statement an agency may request comments on it before the decision is finally made. In any case other agencies or persons may make comments before the final decision unless a different time is provided under § 1506.10.

§ 1503.2   Duty to comment.

Federal agencies with jurisdiction by law or special expertise with respect to any environmental impact involved or which are authorized to develop and enforce environmental standards shall comment on statements within their jurisdiction, expertise, or authority. A Federal agency may (and a cooperating agency that is satisfied that its views are adequately reflected in the environmental impact statement would) reply that it has no comment.

§ 1503.3   Specificity of comments.

Comments on an environmental impact statement or on a proposed action shall be as specific as possible and may address either the adequacy of the statement or the merits of the alternatives discussed or both. When a commenting agency criticizes a lead agency's predictive methodology, the commenting agency should describe the alternative methodology which it prefers and why.

§ 1503.4 Response to comments.

(a) An agency preparing a final environmental impact statement shall assess and consider comments both individually and collectively, and shall respond by one or more of the means listed below specifying its response in the final statement. Possible responses are to:

(1) Modify the proposed action.

(2) Develop and evaluate alternatives not previously given serious consideration by the agency.

(3) Supplement, improve, or modify its analyses.

(4) Make factual corrections.

(5) Explain why the comments do not warrant further agency response, citing the sources, authorities, or reasons which support the agency's position and, if appropriate, indicate those circumstances which would trigger agency reappraisal or further response.

(b) All substantive comments received on the draft statement (or summaries thereof where the response has been exceptionally voluminous), should be attached to the final statement whether or not the comment is thought to merit individual discussion by the agency in the text of the statement.

(c) If changes are minor and are confined to the responses described in paragraphs (a)(4) and (5) of this section, agencies may write them on errata sheets and attach them to the statement instead of rewriting the draft statement. In such cases only the comments, the responses, and the changes and not the final statement need be circulated (§ 1502.19). The entire document with a new cover sheet shall be filed as the final statement (§ 1506.9).

PART 1504—PREDECISION REFERRALS TO THE COUNCIL OF PROPOSED FEDERAL ACTIONS

FOUND TO BE ENVIRONMENTALLY UNSATISFACTORY

Sec.
1504.1 Purpose.
1504.2 Criteria for Referral.
1504.3 Procedure for Referrals and Response.

AUTHORITY: NEPA, the Environmental Quality Improvement Act of 1970, as amended (42 U.S.C. 4371 et seq.), Section 309 of the Clean Air Act, as amended (42 U.S.C. 1857h-7), and Executive Order 11514, Protection and Enhancement of Environmental Quality (March 5, 1970, as amended by Executive Order 11991, May 24, 1977).

§ 1504.1 Purpose.

(a) This part establishes procedures for referring to the Council Federal interagency disagreements concerning proposed major Federal actions that might cause unsatisfactory environmental effects. It provides means for early resolution of such disagreements.

(b) Under section 309 of the Clean Air Act (42 U.S.C. 7609), the Administrator of the Environmental Protection Agency is directed to review and comment publicly on the environmental impacts of Federal activities, including actions for which environmental impact statements are prepared. If after this review the Administrator determines that the matter is "unsatisfactory from the standpoint of public health or welfare or environmental quality," section 309 directs that the matter be referred to the Council (hereafter "environmental referrals").

(c) Under section 102(2)(C) of the Act other Federal agencies may make similar reviews of environmental impact statements, including judgments on the acceptability of anticipated environmental impacts. These reviews must be made available to the President, the Council and the public.

§ 1504.2 Criteria for referral.

Environmental referrals should only be made to the Council after concerted, timely (as early as possible in the process), but unsuccessful attempts to resolve differences with the lead agency. In determining what environmental objections to the matter are appropriate to refer to the Council, an agency should weigh potential adverse environmental impacts, considering:

(a) Possible violation of national environmental standards or policies.

(b) Severity.

(c) Geographical scope.

(d) Duration.

(e) Importance as precedents.

(f) Availability of environmentally preferable alternatives.

§ 1504.3 Procedure for referrals and response.

(a) A Federal agency making the referral to the Council shall:

(1) Advise the lead agency at the earliest possible time that it intends to refer a matter to the Council unless a satisfactory agreement is reached.

(2) Include such advice in the referring agency's comments on the draft environmental impact statement, except when the statement does not contain adequate information to permit an assessment of the matter's environmental acceptability.

(3) Identify any essential information that is lacking and request that it be made available at the earliest possible time.

(4) Send copies of such advice to the Council.

(b) The referring agency shall deliver its referral to the Council not later than twenty-five (25) days after the final environmental impact statement has been made available to the Environmental Protection Agency, commenting agencies, and the public. Except when an extension of this period has been granted by the lead agency, the council will not accept a referral after that date.

(c) The referral shall consist of:

(1) A copy of the letter signed by the head of the referring agency and delivered to the lead agency informing the lead agency of the referral and the reasons for it, and requesting that no action be taken to implement the matter until the Council acts upon the referral. The letter shall include a copy of the statement referred to in § 1504.3(c)(2) below.

(2) A statement supported by factual evidence leading to the conclusion that the matter is unsatisfactory from the standpoint of public health or welfare or environmental quality. The statement shall:

(i) Identify any material facts in controversy and incorporate (by reference if appropriate) agreed upon facts,

(ii) Identify any existing environmental requirements or policies which would be violated by the matter,

(iii) Present the reasons the referring agency believes the matter is environmentally unsatisfactory,

(iv) Contain a finding by the agency whether the issue raised is one of national importance because of the threat to national environmental resources or policies or for some other reason,

(v) Review the steps taken by the referring agency to bring its concerns to the attention of the lead agency at the earliest possible time, and

(vi) Give the referring agency's recommendations as to what mitigation alternative, further study, or other course of action (including abandonment of the matter) are necessary to remedy the situation.

(d) Not later than twenty-five (25) days after the referral to the Council, the lead agency may deliver a response to the Council and the referring agency. If the lead agency requests more time and gives assurance that the matter will not go forward in the interim, the Council may grant an extension. The response shall:

(1) Address fully the issues raised in the referral.

(2) Be supported by evidence.

(3) Give the lead agency's response to the referring agency's recommendations.

(e) Not later than twenty-five (25) days after receipt of both the referral and any response or upon being informed that there will be no response (unless the lead agency agrees to a longer time), the Council may take one or more of the following actions:

(1) Conclude that the process of referral and response has successfully resolved the problem.

(2) Initiate discussions with the agencies with the objective of mediation with referring and lead agencies.

(3) Hold public meetings or hearings to obtain additional views and information.

(4) Determine that the issue is not one of national importance and request the referring and lead agencies to pursue their decision process.

(5) Determine that the issue should be further negotiated by the referring and lead agencies and is not appropriate for Council consideration until one or more heads of agencies report to the Council that the agencies' disagreements are irreconcilable.

(6) Publish its findings and recom-

mendations (including where appropriate a finding that the submitted evidence does not support the position of an agency).

(7) When appropriate, submit the referral and the response together with the Council's recommendation to the President for action.

PART 1505—NEPA AND AGENCY DECISIONMAKING

Sec.
1505.1 Agency decisionmaking procedures.
1505.2 Record of decision in those cases requiring environmental impact statements.
1505.3 Implementing the decision.

AUTHORITY: NEPA, the Environmental Quality Improvement Act of 1970, as amended (42 U.S.C. 4371 et seq.), section 309 of the Clean Air Act, as amended (42 U.S.C. 1857h-7), and Executive Order 11514, Protection and Enhancement of Environmental Quality (March 5, 1970, as amended by Executive Order 11991, May 24, 1977).

§ 1501.1 Agency decisionmaking procedures.

Agencies shall adopt procedures (§ 1507.3) to ensure that decisions are made in accordance with the policies and purposes of the Act. Such procedures shall include but not be limited to:

(a) Implementing procedures under section 102(2) to achieve the requirements of sections 101 and 102(1).

(b) Designating the major decison points for the agency's principal programs likely to have a significant effect on the human environment and assuring that the NEPA process corresponds with them.

(c) Requiring that relevant environmental documents, comments, and responses be part of the record in formal rulemaking or adjudicatory proceedings.

(d) Requiring that relevant environmental documents, comments, and responses accompany the proposal through existing agency review process so that agency officials use the statement in making decisions.

(e) Requiring that the alternatives considered by the decision maker are encompassed by the range of alternatives discussed in the relevant environmental documents and that the decisionmaker consider the alternatives described in the environmental impact statement. If another decision document accompanies the relevant environmental documents to the decisionmaker, agencies are encouraged to make available to the public before the decision is made any part of that document that relates to the comparison of alternatives.

§ 1505.2 Record of decision in those cases requiring environmental impact statements.

At the same time of its decision (or, if appropriate, its recommendation to Congress) each agency shall prepare a concise public record of decision. The record, which may be integrated into any other record prepared by the agency, including that required by OMB Circular A-95, part I, sections 6 (c) and (d), and part II, section 5(b)(4), shall state:

(a) What the decision was.

(b) If an alternative other than those designated pursuant to § 1502.14(e) has been selected, the reasons why other specific considerations of national policy overrode those alternatives.

(c) Whether all practicable means to avoid or minimize environmental harm have been adopted, and if not, why they were not. For any mitigation adopted, a monitoring and enforcement program where applicable shall be adopted and summarized.

§ 1505.3 Implementing the decision.

Agencies may provide for monitoring to assure that their decisions are carried out and should do so in important cases. Mitigation (§ 1505.2(c)) and other conditions established in or during the review of the environmental impact statement and committed as part of the decision shall be implemented by the appropriate agency. The lead agency shall:

(a) Include appropriate conditions in grants, permits or other approvals.

(b) Condition funding of actions on mitigation.

(c) Upon request, inform cooperating or commenting agencies on progress in carrying out mitigation measures proposed by any such agency and adopted by the agency making the decision.

(d) Upon request, make available to the public the results of relevant monitoring.

PART 1506—OTHER REQUIREMENTS OF NEPA

Sec.

AUTHORITY: NEPA, the Environmental Quality Improvement Act of 1970, as amended (42 U.S.C. 4371 et seq.), Section 309 of the Clean Air Act, as amended (42 U.S.C. 1857h-7), and Executive Order 11514, Protection and Enhancement of Environmental Quality (March 5, 1970, as amended by Executive Order 11991, May 24, 1977).

§ 1506.1 Limitations on actions during NEPA process.

(a) Until an agency issues a record of decision as provided in § 1505.2 (except as provided in subsection (c)), no action concerning the proposal shall be taken which would:

(1) Have an adverse environmental impact; or

(2) Limit the choice of reasonable alternatives.

(b) If any agency is considering an application from a non-Federal entity, and is aware that the applicant is planning to take an action within the agency's jurisdiction that would meet either of the criteria in § 1506.1(a), then the agency shall promptly notify the applicant that the agency will take appropriate action to insure that the objectives and procedures of NEPA are achieved.

(c) While work on a required program environmental impact statement is in progress and the action is not covered by an existing program statement, agencies shall not undertake in the interim any major Federal action which may significantly affect the quality of the human environment and which is covered by the program unless such action:

(1) Is justified independently of the program;

(2) Will not prejudice the ultimate decision on the program. Interim action prejudices the ultimate decision on the program when it tends to determine subsequent development or limit alternatives; and

(3) Is itself accompanied by an adequate environmental impact statement.

§ 1506.2 Elimination of duplication with State and local procedures.

(a) Agencies authorized by law to cooperate with State agencies of statewide jurisdiction pursuant to section 102(2)(D) of the Act may do so.

(b) Agencies shall cooperate with State and local agencies to the fullest extent possible to reduce duplication in NEPA and comparable State and local requirements, unless they are specifically barred from doing so by some other law. Except where an agency is proceeding in the manner specified by paragraph (a) of this section, such cooperation shall to the fullest extent possible include:

(1) Joint planning processes.

(2) Joint environmental research and studies.

(3) Joint public hearings (except where otherwise provided by statute).

(4) Joint environmental assessments and joint environmental impact statements. In such cases one or more Federal agencies and one or more State or local agencies shall be joint lead agencies. Where State laws or local ordinances have environmental impact statement requirements in addition to but not in conflict with those in NEPA, Federal agencies shall cooperate in fulfilling the requirements of those as well as Federal laws so that one document will comply with all applicable laws.

(c) To better integrate environmental impact statements into state or local planning processes, statements shall discuss any inconsistency of a proposed action with any approved State or local plan and laws (whether or not federally sanctioned).

§ 1506.3 Adoption.

(a) An agency may adopt a Federal draft or final environmental impact statement or portion thereof provided that the agency treats the statement as a draft and recirculates it (except as provided below in paragraph (b) of this section): *And provided,* That the statement or portions thereof meets the standards for an adequate draft statement under these regulations.

(b) A cooperating agency may adopt

without recirculating the environmental impact statement of a lead agency when, after an independent review of the statement, the cooperating agency concludes that its comments and suggestions have been satisfied.

(c) When an agency adopts a statement which is not final within the agency that prepared it, or when the action it assesses is the subject of a referral under part 1504, or when the statement's adequacy is the subject of a judicial action which is not final, the agency shall so specify.

§ 1506.4 Combining documents.

Any environmental document in compliance with NEPA may be combined with any other agency document to reduce duplication and paperwork.

§ 1506.5 Agency responsibility.

(a) If an agency relies on an applicant to submit initial environmental information, the agency should assist the applicant by outlining the types of information required. In all cases, the agency should make its own evaluation of the environmental issues and take responsibility for the scope and content of environmental assessments.

(b) Except as provided in §§ 1506.2 and 1506.3 any environmental impact statement prepared pursuant to the requirements of NEPA shall be prepared directly by or under contract to the lead agency or where appropriate under § 1501.6(b), a cooperating agency. In the case of such contract it is the intent of these regulations that the contractor be chosen solely by the lead agency or by the lead agency in cooperation with cooperating agencies or where appropriate by a cooperating agency to avoid any conflict of interest. Contractors shall execute a disclosure statement prepared by the lead agency or where appropriate the cooperating agency specifying that they have no financial or other interest in the outcome of the project. If the document is prepared by contract, the responsible Federal official shall furnish guidance and participate in the preparation and shall independently evaluate the statement prior to its approval. Nothing in this section is intended to prohibit any agency from requesting any person to submit information to it or any person from submitting information to any agency.

§ 1506.6 Public involvement.

Agencies shall: (a) Make diligent effort to involve the public in preparing and implementing their NEPA procedures.

(b) Provide public notice of NEPA-related hearings, meetings, and the availability of environmental documents by means calculated to inform those persons and agencies who may be interested or affected.

(1) In all cases the agency shall mail notice to those who have requested it on an individual action.

(2) In the case of an action with effects of national concern such notice shall include publication in the FEDERAL REGISTER and notice by mail to national organizations with interest in the matter and may include listing in the 102 Monitor.

(3) In the case of an action with effects primarily of local concern the notice may include:

(i) Notice to State and local agencies pursuant to OMB Circular A-95.

(ii) Following the affected State's public notice procedures for comparable actions.

(iii) Publication in local newspapers (in papers of general circulation rather than legal papers).

(iv) Notice through other local media.

(v) Notice to potentially interested community organizations including small business associations.

(vi) Publication in newsletters that may be expected to reach potentially interested persons.

(vii) Direct mailing to owners and occupants of nearby or affected property.

(viii) Posting of notice on and off site in the area where the action is to be located.

(c) Hold or sponsor public hearings or public meetings whenever appropriate. Criteria shall include whether there is:

(1) Substantial environmental controversy concerning the proposed action or substantial interest in holding the hearing.

(2) A request for a hearing by another agency with jurisdiction over the action supported by reasons why a hearing will be helpful.

(d) Solicit appropriate information from the public.

(e) Explain in its procedures where

interested persons can get information or status reports on environmental impact statements and other elements of the NEPA process.

(f) Make environmental impact statements, the comments received, and any underlying documents available to the public pursuant to the provisions of the Freedom of Information Act (5 U.S.C. 552), without regard to the exclusion of intra- or interagency memoranda where such memoranda transmit comments of Federal agencies on the environmental impact of the proposed action.

§ 1506.7 Further guidance.

The Council may provide further guidance concerning NEPA and its procedures including:

(a) A handbook which the Council may supplement from time to time which shall in plain language provide guidance and instructions concerning the application of NEPA and these regulations.

(b) Publication of the Council's Memoranda to Heads of Agencies.

(c) In conjunction with the Environmental Protection Agency and the publication of the 102 Monitor, notice of:

(1) Research activities;

(2) Meetings and conferences related to NEPA; and

(3) Successful and innovative procedures used by agencies to implement NEPA.

§ 1506.8 Proposals for legislation.

The NEPA process for proposals for legislation (§ 1508.16) significantly affecting the quality of the human environment shall be integrated with the legislative process of the Congress. A legislative environmental impact statement is the detailed statement required by law which shall accompany proposed legislation to the Congress. Preparation of a legislative environmental impact statement shall include consultation with appropriate agencies (which may be pursuant to OMB Circular A-19) and conform with the requirements of these regulations except as follows:

(a) There need not be a scoping process.

(b) The legislative statement shall otherwise be treated in the same manner as a draft statement except as further specified. There need not be a

final environmental impact statement: *Provided,* That when any of the following conditions exist both the draft and final environmental impact statement on the legislative proposal shall be prepared and circulated as provided by sections 1503.1 and 1506.10.

(1) A Congressional Committee with jurisdiction over the proposal has a rule requiring both draft and final environmental impact statements.

(2) The proposal results from a study process required by statute (such as those required by the Wild and Scenic Rivers Act (16 U.S.C. 1271 et seq.) and the Wilderness Act (16 U.S.C. et seq.)).

(3) Legislative approval is sought for Federal or federally assisted construction or other projects which the agency recommends be located at specific geographic locations. For proposals requiring an environmental impact statement for the acquisition of space by the General Services Administration, a draft statement shall accompany the Prospectus or the 11(b) Report of Building or the 11(b) Report of Building Project Surveys to the Congress, and a final statement shall be completed before site acquisition.

(4) The agency decides to prepare draft and final statements.

(c) Comments on the legislative statement shall be given to the lead agency which shall forward them along with its own responses to the Congressional committees with jurisdiction.

(d) The Environmental Protection Agency may reduce the period for review required by § 1506.10 to insure that comments and responses are received by the appropriate Congressional committee prior to hearings on the proposal.

§ 1506.9 Filing requirements.

Environmental impact statements together with comments and responses shall be filed with the Environmental Protection Agency, attention Office of Federal Activities (A-104), 401 M Street SW., Washington, D.C. 20460. Statements shall be filed with EPA no earlier than they are also transmitted to commenting agencies and the public. EPA shall deliver one copy of each statement to the Council, which shall satisfy the requirement of availability to the President.

§ 1506.10 Timing of agency action.

(a) No decision on the proposed action shall be made or recorded under § 1505.2 by a Federal agency until the later of the following dates:

(1) Ninety (90) days after publication of the notice described in paragraph (d) of this section for a draft environmental impact statement.

(2) Thirty (30) days after publication of the notice described in paragraph (d) of this section for a final environmental impact statement.

Provided, That when an agency has formally established an internal appeal process, through which agencies or the public may take appeals and make their views known after preparation of the final environmental impact statement, and which provides a real opportunity to alter the decision, an administratively reviewable decision in the proposed action may be made after publication of the notice described in paragraph (d) of this section for a final environmental impact statement. This means that the period for appeal and the period prescribed by paragraph (a)(2) of this section may run concurrently. In such a case the environmental impact statement shall explain the timing and the public's right of appeal.

Provided further, That when an agency's primary purpose is the protection of public health and safety, the agency may, with the approval of the Council, adopt procedures under § 1507.3 providing for a finding to be published in the FEDERAL REGISTER that it is necessary to waive the time requirement specified in paragraph (a)(2) of this section to preserve public health and safety.

Provided further, That when an agency's primary purpose is the protection of public health and safety and when that agency publishes proposed rules in the FEDERAL REGISTER for a period of public review prescribed by a statute the agency administers, that time period and the period prescribed under paragraph (a)(2) of this section may run concurrently.

(b) If the final environmental impact statement is filed within ninety (90) days after a draft environmental impact statement is filed with the Environmental Protection Agency, the minimum thirty (30) day period and the minimum ninety (90) day period may run concurrently.

(c) Subject to paragraph (e) of this section agencies shall allow not less than 45 days for comments on draft statements.

(d) The Environmental Protection Agency shall publish a notice in the FEDERAL REGISTER each week of the environmental impact statements filed with the Environmental Protection Agency the preceding week. The date of publication of this notice shall be the date from which the minimum time periods of this section shall be calculated.

(e) The lead agency may extend prescribed periods. The Environmental Protection Agency may upon a showing by the lead agency of compelling reasons of national policy reduce the prescribed periods and may upon a showing by any other Federal agency of compelling reasons of national policy also extend prescribed periods, but only after consultation with the lead agency. (Also see § 1507.3(d).) If the lead agency does not concur, the matter shall be referred to CEQ for resolution. When the Environmental Protection Agency reduces or extends any period of time it shall notify the Council.

§ 1506.11 Emergencies.

Where emergency circumstances make it necessary to take an action with significant environmental impact without observing the provisions of these regulations, the Federal agency proposing to take the action should consult with the Council about alternative arrangements. Agencies and the Council will limit such waivers to actions necessary to control the immediate impacts of the emergency. Other actions remain subject to NEPA review.

§ 1506.12 Effective date.

The effective date of these regulations is eight months after their final publication in the FEDERAL REGISTER.

(a) These regulations shall apply to the fullest extent practicable to ongoing activities and environmental documents begun before the effective date. These regulations do not apply to an environmental impact statement if the draft statement was filed before the effective date of these regulations. No completed environmental documents need be redone by reason of these regulations. Until these regulations are applicable, the Council's guidelines published in the FEDERAL REGISTER of August 1, 1973, shall continue to be applicable. In cases where these regulations are applicable the guidelines are superseded. However, nothing shall prevent an agency from proceeding under these regulations at an earlier time.

(b) NEPA shall continue to be applicable to actions begun before January 1, 1970, to the fullest extent possible.

PART 1507—AGENCY COMPLIANCE

Sec.
1507.1 Compliance.
1507.2 Agency Capability to Comply.
1507.3 Agency Procedures.

AUTHORITY: NEPA, the Environmental Quality Improvement Act of 1970, as amended (42 U.S.C. 4371 et seq.), Section 309 of the Clean Air Act, as amended (42 U.S.C. 1857h-7), and Executive Order 11514, Protection and Enhancement of Environmental Quality (March 5, 1970, as amended by Executive Order 11991, May 24, 1977).

§ 1507.1 Compliance.

All agencies of the Federal Government shall comply with these regulations. It is the intent of these regulations to allow each agency flexibility in adapting its implementing procedures authorized by § 1507.3 to the requirements of other applicable laws.

§ 1507.2 Agency capability to comply.

Each agency shall be capable (in terms of personnel and other resources) of complying with the requirements enumerated below. Such compliance may include use of other's resources, but the using agency shall itself have sufficient capability, at minimum, to evaluate what others do for it. Agencies shall:

(a) Fulfill the requirements of Sec. 102(2)(A) of the Act to utilize a systematic, interdisciplinary approach which will insure the integrated use of the natural and social sciences and the environmental design arts in planning and in decisionmaking which may have an impact on the human environment. Agencies shall designate a person to be responsible for overall review of agency NEPA compliance.

(b) Identify methods and procedures required by Sec. 102(2)(B) to insure

that presently unquantified environmental amenities and values may be given appropriate consideration.

(c) Prepare adequate environmental impact statements pursuant to Sec. 102(2)(C) and comment on statements in the areas where the agency has jurisdiction by law or special expertise or is authorized to develop and enforce environmental standards.

(d) Study, develop, and describe alternatives to recommended courses of action in any proposal which involves unresolved conflicts concerning alternative uses of available resources. This requirement of Sec. 102(2)(E) extends to all such proposals, not just the more limited scope of Sec. 102(2) (C)(iii) where the discussion of alternatives is confined to impact statements.

(e) Comply with the requirements of Sec. 102(2)(H) that the agency initiate and utilize ecological information in the planning and development of resource-oriented projects.

(f) Fulfill the requirements of sections 102(2)(F), 102(2)(G), and 102(2)(I), of the Act and of Executive Order 11514, Protection and Enhancement of Environmental Quality, Sec. 2.

§ 1507.3 Agency procedures.

(a) Not later than eight months after publication of these regulations as finally adopted in the FEDERAL REGISTER, or five months after the establishment of an agency, whichever shall come later, each agency shall as necessary adopt procedures to supplement these regulations. When the agency is a department major subunits are encouraged (with the consent of the department) to adopt their own procedures. Such procedures shall not paraphrase these regulations. They shall confine themselves to implementing procedures. Each agency shall consult with the Council while developing its procedures and before publishing them in the FEDERAL REGISTER for comment. The procedures shall be adopted only after an opportunity for public review and after review by the Council for conformity with the Act and these regulations. The Council shall complete its review within 30 days. Once in effect they shall be filed with the Council and made readily available to the public. Agencies are encouraged to publish explanatory guidance for these regulations and their own procedures. Agencies shall continue to review their policies and procedures and to revise them as necessary to ensure full compliance with the purposes and provisions of the Act.

(b) Agency procedures shall comply with these regulations except where compliance would be inconsistent with statutory requirements and shall include:

(1) Those procedures required by §§ 1501.2(d), 1502.9(c)(3), 1503.1(c), 1505.1, 1506.6(e), and 1508.4.

(2) Specific criteria for and identification of those typical classes of action:

(i) Which normally do require environmental impact statements.

(ii) Which normally do not require either an environmental impact statement or an environmental assessment (categorical exclusions (§ 1508.4)).

(iii) Which normally require environmental assessments but not necessarily environmental impact statements.

(c) Agency procedures may include specific criteria for providing limited exceptions to the provisions of these regulations for proposed actions that are specifically authorized under criteria established by an Executive Order or statute to be kept secret in the interest of national defense or foreign policy and are in fact properly classified pursuant to such Executive Order or statute. Environmental assessments and environmental impact statements which address classified proposals may be safeguarded and restricted from public dissemination in accordance with agencies' own regulations applicable to classified information. These documents may be organized so that classified portions can be included as annexes, in order that the unclassified portions can be made available to the public.

(d) Agency procedures may provide for periods of time other than those presented in § 1506.10 when necessary to comply with other specific statutory requirements.

PART 1508—TERMINOLOGY AND INDEX

AUTHORITY: NEPA, the Environmental Quality Improvement Act of 1970, as amended (42 U.S.C. 4371 *et seq.*), Section 309 of the Clean Air Act, as amended (42 U.S.C. 1857h-7), and Executive Order 11514, Protection and Enhancement of Environmental Quality (March 5, 1970, as amended by Executive Order 11991, May 24, 1977).

§ 1508.1   Terminology.

The terminology of this part shall be uniform throughout the Federal Government.

§ 1508.2   Act.

"Act" means the National Environmental Policy Act, as amended (42 U.S.C. 4321, et seq.) which is also referred to as "NEPA."

§ 1508.3   Affecting.

"Affecting" means will or may have an effect on.

§ 1508.4   Categorical exclusion.

"Categorical Exclusion" means a category of actions which do not individually or cumulatively have a significant effect on the human environment and which have been found to have no such effect in procedures adopted by a Federal agency in implementation of these regulations (§ 1507.3) and for which, therefore, neither an environmental assessment nor an environmental impact is needed. Any such procedures shall provide for extraordinary circumstances in which a normally excluded action may have a significant environmental effect.

§ 1508.5   Cooperating agency.

"Cooperating Agency" means any

Federal agency other than a lead agency which has jurisdiction by law or special expertise with respect to any environmental impact involved in a proposal (or a reasonable alternative) for legislation or other major Federal action significantly affecting the quality of the human environment. The selection and responsibilities of a cooperating agency are described in § 1501.6. A State or local agency of similar qualifications or, when the effects are on a reservation, an Indian Tribe may by agreement with the lead agency become a cooperating agency.

§ 1508.6 Council.

"Council" means the Council on Environmental Quality established by Title II of the Act.

§ 1508.7 Cumulative impact.

"Cumulative impact" is the impact on the environment which results from the incremental impact of the action when added to other past, present, and reasonably foreseeable future actions regardless of what agency (Federal or non-Federal) or person undertakes such other actions. Cumulative impacts can result from individually minor but collectively significant actions taking place over a period of time.

§ 1508.8 Effects.

"Effects" include:

(a) Direct effects, which are caused by the action and occur at the same time and place.

(b) Indirect effects, which are caused by the action and are later in time or farther removed in distance, but are still reasonably foreseeable. Indirect effects may include growth inducing effects and other effects related to induced changes in the pattern of land use, population density or growth rate, and related effects on air and water and other natural systems, including ecosystems.

Effects and impacts as used in these regulations are synonymous. Effects includes ecological (such as the effects on natural resources and on the components, structures, and functioning of affected ecosystems), economic, social, or health, whether direct, indirect, or cumulative. Effects may also include those resulting from actions which may have both beneficial and detrimental effects, even if on balance the agency believes that the effect will be beneficial.

§ 1508.9 Environmental assessment.

"Environmental Assessment":

(a) Means a public document for which a Federal agency is responsible that serves to:

(1) Briefly provide sufficient evidence and analysis for determining whether to prepare an environmental impact statement or a finding of no significant impact.

(2) Aid an agency's compliance with the Act when no environmental impact statement is necessary.

(3) Facilitate preparation of such a statement when one is necessary.

(b) Shall include brief discussions of the need for the proposal, of alternatives as required by sec. 102(2)(E), of the environmental impacts of the proposed action and alternatives, and a listing of agencies and persons consulted. Most environmental impacts of the proposed action and alternatives, and a listing of agencies and persons consulted. Most environmental assessments do not exceed several pages in length.

§ 1508.10 Environmental document

"Environmental Document" includes the documents specified in §§ 1508.9, 1508.11, 1508.13 and 1508.21.

§ 1508.11 Environmental impact statement

"Environmental Impact Statement" means a detailed written statement as required by Sec. 102(2)(C) of the Act.

§ 1508.12 Federal agency.

"Federal agency" means all agencies of the Federal Government. It does not mean the Congress, the Judiciary, or the President, including the performance of staff functions for the President in his Executive Office.

§ 1508.13 Finding of no significant impact.

"Finding of No Significant Impact" means a document by a Federal agency briefly presenting the reasons why an action, not otherwise excluded (§ 1508.4), will not have a significant effect on the human environment and for which an environmental impact statement therefore will not be prepared. It shall include the environmental assessment or a summary of it and shall note any other environmental documents related to it (§ 1501.7(a)(5)).

§ 1508.14 Human environment.

"Human Environment" shall be interpreted comprehensively to include the natural and physical environment and the interaction of people with that environment. (See the definition of "effects" (§ 1508.8).) This means that exclusively economic or social effects are not intended by themselves to require preparation of an environmental impact statement. When an environmental impact statement is prepared and economic or social and natural or physical environmental effects are interrelated, then the environmental impact statement will discuss all of these effects on the human environment.

§ 1508.15 Lead agency.

"Lead Agency" means the agency or agencies which have prepared or have taken primary responsibility to prepare the environmental impact statement.

§ 1508.16 Legislation.

"Legislation" includes a bill or legislative proposal to Congress developed by or with the significant cooperation and support of a Federal agency, but does not include requests for appropriations.[1] The test for significant cooperation is whether the proposal is in fact predominantly that of the agency rather than another source. Drafting does not by itself constitute significant cooperation. Proposals for legislation include requests for ratification of treaties. Only the agency which has primary responsibility for the subject matter involved will prepare a legislative environmental impact statement.

§ 1508.17 Major Federal action.

"Major Federal action" includes actions with effects that may be major and which are potentially subject to Federal control and responsibility. Major reinforces but does not have a meaning independent of significantly (§ 1508.25). Actions include the circumstance where the responsible officials fail to act and that failure to act is re-

---

[1] The Council in consultation with OMB had been prepared to propose this wording and § 1508.12 for comment. Thereafter *Sierra Club* v. *Andrus* (D.C. Cir. No. 75-1871, May 15, 1978) was decided. We would appreciate comment on the implications of that case for these provisions.

viewable by courts or administrative tribunals under the Administrative Procedure Act or other applicable law as agency action. If a Federal program is delegated or otherwise transferred to State or local government, unless Congress intended otherwise, the Federal agency shall continue to be responsible for compliance with the Act and shall insure the preparation of environmental impact statements if they would be required but for the delegation or transfer. If the Federal agency may legally require the State or local agency to follow an environmental impact statement process, as a condition of the delegation or transfer, it shall do so. If not, the Federal agency shall prepare the statements (except as provided in § 1506.5).

(a) Actions include new and continuing activities, including projects and programs entirely or partly financed, assisted, conducted, regulated, or approved by federal agencies; new or revised agency rules, regulations, plans, policies, or procedures; and legislative proposals (§§ 1506.8, 1508.16). Actions do not include funding assistance solely in the form of general revenue sharing funds, distributed under the State and Local Fiscal Assistance Act of 1972, 31 U.S.C. 1221 et seq., with no Federal agency control over the subsequent use of such funds. Actions do not include bringing civil or criminal enforcement actions.

(b) Federal actions tend to fall within one of the following categories:

(1) Adoption of official policy, such as rules, regulations, and interpretations adopted pursuant to the Administrative Procedure Act, 5 U.S.C. 551 et seq.; treaties and international conventions or agreements; formal documents establishing an agency's policies which will result in or substantially alter agency programs.

(2) Adoption of formal plans, such as official documents prepared or approved by federal agencies which guide or prescribe alternative uses of federal resources, upon which future agency actions will be based.

(3) Adoption of programs, such as a group of concerted actions to implement a specific policy or plan; systematic and connected agency decisions allocating agency resources to implement a specific statutory program or executive directive.

(4) Approval of specific projects, such as construction or management

activities located in a defined geographic area. Projects include actions approved by permit or other regulatory decision as well as federal and federally assisted activities.

§ 1508.18 Matter.

"Matter" includes for purposes of Part 1504:

(a) With respect to the Environmental Protection Agency, any proposed legislation, project, action or regulation as those terms are used in Section 309(a) of the Clean Air Act (42 U.S.C. 7609).

(b) With respect to all other agencies, any proposed major federal action to which Section 102(2)(C) of NEPA applies.

§ 1508.19 Mitigation.

"Mitigation" includes:

(a) Avoiding the impact altogether by not taking a certain action or parts of an action.

(b) Minimizing impacts by limiting the degree or magnitude of the action and its implementation.

(c) Rectifying the impact by repairing, rehabilitating, or restoring the impacted environment.

(d) Reducing or eliminating the impact over time by preservation and maintenance operations during the life of the action.

(e) Compensating for the impact by replacing or providing substitute resources or environments.

§ 1508.20 NEPA process.

"NEPA process" means all measures necessary for compliance with the requirements of Section 2 and Title I of NEPA.

§ 1508.21 Notice of intent.

"Notice of Intent" means a notice that an environmental impact statement will be prepared and considered. The notice shall briefly:

(a) Describe the proposed action and possible alternatives.

(b) Describe the agency's proposed scoping process including whether, when, and where any scoping meeting will be held.

(c) State the name and address of a person within the agency who can answer questions about the proposed action and the environmental impact statement.

§ 1508.22 Proposal.

"Proposal" refers to that stage in the development of an action when an agency subject to the Act has a goal and is actively considering one or more alternative means of accomplishing that goal and the effects can be meaningfully evaluated. Preparation of an environmental impact statement on a proposal should be timed (§ 1502.5) so that the final statement may be completed in time for the statement to be included in any recommendation or report on the proposal. A proposal may exist in fact as well as by agency declaration that one exists.

§ 1508.23 Referring agency.

"Referring agency" means the federal agency which has referred any matter to the Council after a determination that the matter is unsatisfactory from the standpoint of public health or welfare or environmental quality.

§ 1508.24 Scope.

Scope consists of the range of actions, alternatives, and impacts to be considered in an environmental impact statement. The scope of an individual statement may depend on its relationships to other statements (§§ 1502.20 and 1508.26). In scoping environmental impact statements agencies shall consider 3 types of actions, 3 types of alternatives, and 3 types of impacts. They include:

(a) Actions (other than unconnected single actions) which may be:

(1) Connected actions, which means that they are closely related and therefore should be discussed in the same impact statement. Actions are connected if they:

(i) Automatically trigger other actions which may require environmental impact statements.

(ii) Cannot or will not proceed unless other actions are taken previously or simultaneously.

(iii) Are interdependent parts of a larger action and depend on the larger action for their justification.

(2) Cumulative actions, which when viewed with other proposed actions have cumulatively significant impacts and should therefore be discussed in the same impact statement.

(3) Similar actions, which when viewed with other reasonably foreseeable or proposed agency actions, have

similarities that provide a basis for evaluating their environmental consequences together, such as common timing or geography. An agency may wish to analyze these actions in the same impact statement. It should do so when the best way to assess adequately the combined impacts of similar actions or reasonable alternatives to such actions is to treat them in a single impact statement.

(b) Alternatives, which include: (1) No action alternative. (2) Other reasonable courses of actions. (3) Mitigation measures (not in the proposed action).

(c) Impacts, which may be: (1) Direct. (2) Indirect. (3) Cumulative.

§ 1508.25 Significantly.

"Significantly" as used in NEPA requires considerations of both context and intensity:

(a) *Context.* This means that the significance of an action must be analyzed in several contexts such as society as a whole (global, national), the affected region, the affected interests, and the locality. Significant varies with the setting of the proposed action. For instance, in the case of a site-specific action, significance would usually depend upon the effects in the locale rather than in the world as a whole. Both short- and long-term effects are relevant.

(b) *Intensity.* This refers to the severity of impact. Responsible officials must bear in mind that more than one agency may make decisions about partial aspects of a major action. The following should be considered in evaluating intensity:

(1) Impacts that may be both beneficial and adverse. A significant effect may exist even if the Federal agency believes that on balance the effect will be beneficial.

(2) The degree to which the proposed action affects public health or safety.

(3) Unique characteristics of the geographic area such as proximity to historic sites, park lands, prime farm lands, wetlands, wild and scenic rivers, or ecologically critical areas.

(4) The degree to which the effects on the quality of the human environment are likely to be highly controversial.

(5) The degree to which the possible effects on the human environment are highly uncertain or involve unique or unknown risks.

(6) The degree to which the action may establish a precedent for future actions with significant effects or represents a decision in principle about a future consideration.

(7) Whether the action is related to other actions with individually insignificant but cumulatively significant impacts. Significance exists if it is reasonable to anticipate a cumulatively significant impact on the environment. Significance cannot be avoided by terming an action temporary or by breaking it down into small component parts.

(8) Whether the action may have a significant adverse effect on an area or site listed in or eligible for listing in the National Register of Historic Places or may cause loss or destruction of significant scientific, cultural, or historical resources.

(9) Whether the action may have a significant adverse effect on the habitat or an endangered or threatened species that has been determined to be critical under the Endangered Species Act of 1973.

(10) Whether the action threatens a violation of Federal, State, or local law or requirements imposed for the protection of the environment.

§ 1508.26 Tiering.

"Tiering" refers to the coverage of general matters in broader environmental impact statements (such as national program or policy statements) with subsequent narrower statements or environmental analyses (such as regional or basinwide program statements or ultimately site-specific statements) incorporating by reference the general discussions and concentrating solely on the issues specific to the statement subsequently prepared. Tiering is appropriate when the sequence of statements or analyses is:

(a) From a program, plan, or policy environmental impact statement to a program, plan, or policy statement or analysis of lesser scope or to a site-specific statement or analysis.

(b) From an environmental impact statement on a specific action at an early stage (such as need and site selection) to a supplement (which is preferred) or a subsequent statement or analysis at a later stage (such as design detail and environmental mitigation). Tiering in such cases is appropriate when it helps the lead agency to focus on the issues which are ripe for decision and exclude from consideration issues already decided or not yet ripe.

INDEX

[FR Doc. 78-15700 Filed 6-8-78; 8:45 am]

D

Federal Agency
NEPA Regulations

April 1, 1978

The following table contains citations to federal agency regulations which tailor the environmental impact statement process to the particular programs administered by each agency. The regulations were adopted pursuant to the National Environmental Policy Act and the Council on Environmental Quality's guidelines for implementing NEPA.

The guidelines are presently being revised. When the revision is concluded, the regulations in the table may be amended to incorporate changes in the operation of the impact statement process introduced by such revision.

Citations refer to the *Federal Register* (Fed. Reg.) and the *Code of Federal Regulations* (C.F.R.), which are available in most law libraries. Unless otherwise noted, the "date" column refers to the effective date of the regulations. Single copies of the regulations usually may be obtained from the appropriate federal offices listed in Appendix E.

Agency	Date	Current Procedures; Notes
Consumer Products Safety Commission	5/18/77	42 Fed. Reg. 25494 (interim regulations)
Department of Agriculture		
Departmental	5/29/74	39 Fed. Reg. 18678
Agricultural Stabilization and Conservation Service	1/15/75	7 C.F.R. Part 799; 39 Fed. Reg. 43996
Farmer's Home Administration	6/09/76	7 C.F.R. Part 1901, Subpart G; 41 Fed. Reg. 22256 (amended by 41 Fed. Reg. 23186, published on 6/09/76)
Forest Service	10/30/74	39 Fed. Reg. 38244
Rural Electrification Administration	5/20/74	39 Fed. Reg. 23240
Soil Conservation Service	6/22/76	7 C.F.R. Part 650; 39 Fed. Reg. 19648 (amended by 41 Fed. Reg. 24976, published on 6/22/76)
Department of Commerce	2/04/75	40 Fed. Reg. 5175; Departmental Order 216-6
Department of Defense		
U.S. Army Corps of Engineers	4/15/74	33 C.F.R. § 209.410; 39 Fed. Reg. 12737 (See also 33 C.F.R. Part 294)
Department of Energy	2/21/78	43 Fed. Reg. 7232 (Proposed departmental regulations. When adopted, these regulations will supersede those of the Federal Energy Administration [FEA] and the Energy Research and Development Administration [ERDA]. In the interim, regulations found in 10 C.F.R. § 208 [41 Fed. Reg. 4724] apply to former FEA activities, and regulations found in 10 C.F.R. § 790.23 [42 Fed. Reg. 4826] apply to former ERDA activities.)

Agency	Date	Current Procedures; Notes
Department of Health, Education, and Welfare (DHEW)		
Departmental	10/17/73	DHEW General Administration Manual, Chapters 30-10 through 30-16
Food and Drug Administration	5/16/77	21 C.F.R. Part 25; 42 Fed. Reg. 15634 (amended by 42 Fed. Reg. 19986, published on 4/15/77)
Department of Housing and Urban Development		
Departmental	6/11/76	Departmental Handbook 1390.1 (amended by 41 Fed. Reg. 23878, published on 6/11/76)
Community Block Grant Program	7/16/75	24 C.F.R. Part 58; 40 Fed. Reg. 29992
Department of the Interior		
Departmental	9/27/71	36 Fed. Reg. 19343; Department Manual Part 516
Bureau of Mines	2/09/72	37 Fed. Reg. 2895
Bureau of Reclamation	11/23/72	37 Fed. Reg. 24910
Heritage Conservation and Recreation Service (formerly the Bureau of Outdoor Recreation)	3/30/72	37 Fed. Reg. 6501
Department of Transportation (DOT)		
Departmental	9/30/74	39 Fed. Reg. 35232
Federal Aviation Administration (FAA)	6/27/77	FAA Order 1050.1B; 42 Fed. Reg. 32631
	10/21/76	FAA Order 5050.2B; 41 Fed. Reg. 46434
Federal Highway Administration	12/02/74	23 C.F.R. Part 771; 39 Fed. Reg. 41804
Federal Railroad Administration	3/05/76	49 C.F.R. § 255.15; 41 Fed. Reg. 9693

Agency	Date	Current Procedures; Notes
National Highway Traffic Safety Administration	11/04/75	49 C.F.R. Part 520; 40 Fed. Reg. 52395
United States Coast Guard	10/22/75	40 Fed. Reg. 49383
Urban Mass Transportation Administration	2/01/72	DOT Order 5610.1; 37 Fed. Reg. 22692 (amended by 41 Fed. Reg. 41512, published on 9/22/76)
Environmental Protection Agency	4/14/75	40 C.F.R. Part 6; 40 Fed. Reg. 16813 (amended by 42 Fed. Reg. 2450, published on 1/11/77)
General Services Administration (GSA)	4/27/77	GSA Order 1095.1A; 42 Fed. Reg. 24095 (general regulations)
	7/01/75	GSA Order PBS 1095; 40 Fed. Reg. 27733 (specific regulations concerning real property)
	7/23/77	42 Fed. Reg. 48387 (specific regulations concerning acquisition of rental space)
Interstate Commerce Commission	7/07/76	49 C.F.R. Part 1108; 41 Fed. Reg. 27838
Nuclear Regulatory Commission	7/18/74	10 C.F.R. Part 51; 39 Fed. Reg. 26279 (amended by 42 Fed. Reg. 13803, published on 3/14/77; by 42 Fed. Reg. 18387, published on 4/7/77; and by 42 Fed. Reg. 34276, published on 7/5/77)

E

Individuals in Federal Agencies Who Can Supply Information on NEPA

The following table lists, for each major federal agency involved in the preparation or review of impact statements, the appropriate person to contact for assistance. Where available, the name and telephone number of this individual is given, together with his title and mailing address. This information was current as of April 1, 1978. At the end of the appendix is a map that shows the standard geographic areas covered by the regional offices of the federal agencies.

AGENCY	PERSON OR OFFICE TO CONTACT	ADDITIONAL INFORMATION; COMMENTS
Consumer Products Safety Commission (CPSC)	Office of the Secretary Consumer Products Safety Commission 1111 18th Street, N.W. Washington, D.C. 20207 Telephone: (202)634-7700	The Office of the Secretary should be contacted for copies of CPSC's statements and with requests for review of other agency statements. For referral to the author of a specific impact statement issued by CPSC, contact: Mr. Michael Brown Office of the Executive Director Consumer Products Safety Division 5401 Westbard Avenue Bethesda, Maryland 20207 Telephone: (301)492-6550
Department of Agriculture	Mr. Barry Flamm Coordinator of Environmental Quality Activities Office of the Secretary U.S. Department of Agriculture Room 359-A Washington, D.C. 20250 Telephone: (202)447-3965	
Department of Commerce	Dr. Sidney R. Galler Deputy Assistant Secretary for Environmental Affairs Department of Commerce Washington, D.C. 20230 Telephone: (202)377-4335	
Department of Defense, U.S. Army Corps of Engineers	Dr. C. Grant Ash Office of Environmental Policy Development ATTN: DAEN-CWR-P Office of the Chief of Engineers Civil Works Directorate U.S. Army Corps of Engineers 1000 Independence Avenue, S.W. Washington, D.C. 20314 Telephone: (202)693-6795	This office is a central referral office. Requests which it receives for review of other agency statements will be forwarded to the district engineer responsible for the area of the proposed site. Requests for review of a statement also may be sent directly to the appropriate district engineer.

AGENCY	PERSON OR OFFICE TO CONTACT	ADDITIONAL INFORMATION; COMMENTS
Department of Energy	Dr. Robert Stern, Director Room 7119 Environmental Impact Division Department of Energy Old Post Office Building 12th and Pennsylvania Avenue, N.W. Washington, D.C. 20461 Telephone: (202)566-9760	
Department of Health, Education, and Welfare (DHEW)	Mr. Charles Custard Office of the Assistant Secretary for Administration and Management Department of Health, Education, and Welfare Room 524F, South Portal Building Washington, D.C. 20201 Telephone: (202)245-7243	The national office coordinates the impact statement activities of DHEW's regional offices. All requests for review of another agency's impact statement should be sent to this office; they will be referred to the appropriate region. Requests for copies of DHEW's impact statements on legislation, national program proposals, and national policy issues should be sent to the national office. Requests for copies of all other DHEW impact statements should be sent to the regional office for the area in which the project is planned. Region I: Donald Branum Regional Environmental Officer Office of Environmental Affairs U.S. Department of Health, Education, and Welfare 2403 John F. Kennedy Center Boston, Massachusetts 02203 Telephone: (617)223-6837

AGENCY	PERSON OR OFFICE TO CONTACT	ADDITIONAL INFORMATION; COMMENTS
Department of Health, Education, and Welfare (cont)		Region II: Frank Trentacosti Regional Engineer U.S. Department of Health, Education, and Welfare 3309 Federal Building 26 Federal Plaza New York, New York 10007 Telephone: (212)264-4483 Region III: Martin Keely Office of Intergovernmental Affairs U.S. Department of Health, Education, and Welfare P.O. Box 13716 Philadelphia, Pennsylvania 19101 Telephone: (215)596-6476 Region IV: Phillip P. Sayre Regional Environmental Officer Office of Intergovernmental Affairs U.S. Department of Health, Education, and Welfare Room 434 50 Seventh Street, N.E. Atlanta, George 30323 Telephone: (404)257-3079 Region V: Melvin H. Fisher Regional Environmental Officer Office of Intergovernmental Affairs U.S. Department of Health, Education, and Welfare 16th Floor 300 S. Wacker Drive Chicago, Illinois 60606 Telephone: (312)353-8874

AGENCY	PERSON OR OFFICE TO CONTACT	ADDITIONAL INFORMATION; COMMENTS
Department of Health, Education, and Welfare (cont)		Region VI: D. Dean Blue Regional Environmental Officer U.S. Department of Health, Education, and Welfare Suite 1635 1200 Main Tower Building Dallas, Texas 75202 Telephone: (214)729-3491 Region VII: William Henderson Regional Environmental Officer Office of Regional Director U.S. Department of Health, Education, and Welfare Room 616 601 East 12th Street Kansas City, Missouri 64106 Telephone: (816)758-5012 Region VIII: Clell Elwood Assistant Regional Director for Intergovernmental Affairs U.S. Department of Health, Education, and Welfare 10th Floor 1961 Stout Street Denver, Colorado 80202 Telephone: (303)327-2831 Region IX: John Knochenhauer Regional Environmental Officer Office of Environmental Affairs U.S. Department of Health, Education, and Welfare Room 423 50 United Nations Plaza San Francisco, California 94102 Telephone: (415)556-2687

AGENCY	PERSON OR OFFICE TO CONTACT	ADDITIONAL INFORMATION; COMMENTS
Department of Health, Education, and Welfare (cont)		Region X: David Miller Regional Environmental Officer U.S. Department of Health, Education, and Welfare Mail Stop 613 Arcade Plaza Building Seattle, Washington 98101 Telephone: (206)399-1290
Department of Housing and Urban Development (HUD)		
Community Block Grant Program	Impact statements should be requested from the city in which the proposed project is located by calling either the mayor's office or the Department of Community Development.	
Departmental	Mr. Richard H. Broun Director Office of Environmental Quality Department of Housing and Urban Development 415 7th Street, S.W. Washington, D.C. 20410 Telephone: (202)755-6308	This office should be contacted for copies of HUD's impact statements on legislation, regulations, national program proposals and major policy issues. For copies of impact statements on other HUD activities and all other requests, contact the regional office for the area in which the project is planned. Region I: David Prescott Environmental Clearance Officer U.S. Department of Housing and Urban Development 800 John F. Kennedy Federal Building Boston, Massachusetts 02203 Telephone: (617)223-4066

AGENCY	PERSON OR OFFICE TO CONTACT	ADDITIONAL INFORMATION; COMMENTS
Department of Housing and Urban Development (cont)		Region II: William J. Davis Environmental Clearance Officer U.S. Department of Housing and Urban Development 26 Federal Plaza New York, New York 10007 Telephone: (212)264-8068 Region III: Robert Dinney Environmental Clearance Officer U.S. Department of Housing and Urban Development Curtis Building, Sixth and Walnut Streets Philadelphia, Pennsylvania 19106 Telephone: (215)597-2560 Region IV: Ivar Iverson Environmental Clearance Officer U.S. Department of Housing and Urban Development 1371 Peachtree Street, N.E. Atlanta, Georgia 30323 Telephone: (404)881-5585 Region V: Richard Kaiser Environmental Clearance Officer U.S. Department of Housing and Urban Development 300 South Wacker Drive Chicago, Illinois 60601 Telephone: (312)353-5680 Region VI: Travis Miller Environmental Clearance Officer U.S. Department of Housing and Urban Development

AGENCY	PERSON OR OFFICE TO CONTACT	ADDITIONAL INFORMATION; COMMENTS
Department of Housing and Urban Development (cont)		Earle Cadell Building 1100 Commerce Street Dallas, Texas 75202 Telephone: (817)749-7401 Region VII: Emil Huber Environmental Clearance Officer U.S. Department of Housing and Urban Development Room 300 911 Walnut Street Kansas City, Missouri 64106 Telephone: (816)374-2661 Region VIII: Robert Matuschek Environmental Clearance Officer U.S. Department of Housing and Urban Development Executive Tower Building 1405 Curtis Street Denver, Colorado 80202 Telephone: (303)837-4513 Region IX: Elizabeth Tapscott Environmental Clearance Officer U.S. Department of Housing and Urban Development 450 Golden Gate Avenue Box 36003 San Francisco, California 94102 Telephone: (415)556-4752 Region X: Robert Scalia Environmental Clearance Officer U.S. Department of Housing and Urban Development 3003 Arcade Plaza Building 1321 Second Avenue Seattle, Washington 98101 Telephone: (206)442-5414

AGENCY	PERSON OR OFFICE TO CONTACT	ADDITIONAL INFORMATION; COMMENTS
Department of the Interior	Mr. Bruce Blanchard Director Environmental Project Review Room 7260 Department of the Interior Washington, D.C. 20240 Telephone: (202)343-3891	The Office of Environmental Project Review acts as the Department of the Interior's central office. Requests sent to this office will be referred to the appropriate departmental subdivision for response.
Department of Transportation		
Departmental	Mr. Martin Convisser Office of Environmental Affairs Office of the Assistant Secretary for Environment, Safety and Consumer Affairs U.S. Department of Transportation 400 7th Street, S.W. Washington, D.C. 20590 Telephone: (202)426-4357	This office is to be contacted for copies of the department's impact statements on legislation, regulations, national program proposals, and major policy issues. For copies of statements on other activities and all other requests, contact the regional office for the area in which the project is planned.
		Region I: Secretarial Representative U.S. Department of Transportation Transportation Systems Center 55 Broadway Cambridge, Massachusetts 02142 Telephone: (617)494-2709
		Region II: Secretarial Representative U.S. Department of Transportation Room 1811 26 Federal Plaza New York, New York 10007 Telephone: (212)264-2672
		Region III: Secretarial Representative U.S. Department of Transportation

AGENCY	PERSON OR OFFICE TO CONTACT	ADDITIONAL INFORMATION; COMMENTS
Department of Transportation Departmental (cont)		Suite 1000 434 Walnut Street Philadelphia, Pennsylvania 19106 Telephone: (215)597-9430

Region IV:
Secretarial Representative
U.S. Department of
   Transportation
Suite 515
1720 Peachtree Road, N.W.
Atlanta, Georgia 30309
Telephone: (404)526-3738

Region V:
Secretarial Representative
U.S. Department of
   Transportation
17th Floor
300 S. Wacker Drive
Chicago, Illinois 60606
Telephone: (312)353-4000

Region VI:
Secretarial Representative
U.S. Department of
   Transportation
9-C-18 Federal Center
1100 Commerce Street
Dallas, Texas 75202
Telephone: (214)749-1851

Region VII:
Secretarial Representative
U.S. Department of
   Transportation
Room 634
601 E. 12th Street
Kansas City, Missouri 64106
Telephone: (816)374-5801

Region VIII:
Secretarial Representative
U.S. Department of
   Transportation
1822 Prudential Plaza
1050 17th Street
Denver, Colorado 80225
Telephone: (303)837-3242

AGENCY	PERSON OR OFFICE TO CONTACT	ADDITIONAL INFORMATION; COMMENTS
Department of Transportation		
Departmental (cont)		Region IX: Secretarial Representative U.S. Department of Transportation 450 Golden Gate Avenue Box 36133 San Francisco, California 94102 Telephone: (415)556-5961 Region X: Secretarial Representative U.S. Department of Transportation 3112 Federal Building 915 Second Avenue Seattle, Washington 98174 Telephone: (206)442-0590
U.S. Coast Guard	Office of Marine Environment and Systems U.S. Coast Guard 400 7th Street, S.W. Washington, D.C. 20591 Telephone: (202)426-2007	
Federal Aviation Administration (FAA)	Office of Environmental Quality Federal Aviation Administration 800 Independence Avenue, S.W. Washington, D.C. 20591 Telephone: (202)426-8406	New England Region: Office of the Regional Director Federal Aviation Administration 12 New England Executive Park 154 Middlesex Street Burlington, Massachusetts 01803 Telephone: (617)273-7244 Eastern Region: Office of the Regional Director Federal Aviation Administration Federal Building

AGENCY	PERSON OR OFFICE TO CONTACT	ADDITIONAL INFORMATION; COMMENTS
Federal Aviation Administration (FAA) (cont)		JFK International Airport Jamaica, New York 11430 Telephone: (212)995-3333

Southern Region:
Office of the Regional
   Director
Federal Aviation
   Administration
P.O. Box 20636
Atlanta, Georgia 30320
Telephone: (404)526-7222

Great Lakes Region:
Office of the Regional
   Director
Federal Aviation
   Administration
2300 East Devon
Des Plaines, Illinois 60018
Telephone: (312)694-4500

Southwest Region:
Office of the Regional
   Director
Federal Aviation
   Administration
P.O. Box 1689
Fort Worth, Texas 76101
Telephone: (817)624-4911

Central Region:
Office of the Regional
   Director
Federal Aviation
   Administration
601 E. 12th Street
Kansas City, Missouri 64106
Telephone: (816)374-5626

Rocky Mountain Region:
Office of the Regional
   Director
Federal Aviation
   Administration
10455 East 25th Street
Aurora, Colorado 80010
Telephone: (303)837-3646

AGENCY	PERSON OR OFFICE TO CONTACT	ADDITIONAL INFORMATION; COMMENTS
Federal Aviation Administration (FAA) (cont)		Western Region: Office of the Regional Director Federal Aviation Administration P.O. Box 92007 WorldWay Postal Center Los Angeles, California 90009 Telephone: (213)536-6435
		Northwest Region: Office of the Regional Director Federal Aviation Administration FAA Building Boeing Field Seattle, Washington 98108 Telephone: (206)767-2780
Federal Highway Administration	Office of Environmental Policy Federal Highway Administration 400 7th Street, S.W. Washington, D.C. 20591 Telephone: (202)426-0351	Regions I and II: Regional Administrator Federal Highway Administration 729 Federal Building Clinton Avenue and North Pearl Street Albany, New York 12207 Telephone: (518)472-6476
		Region III: Regional Administrator Federal Highway Administration 1621 George H. Fallon Federal Office Bldg. 31 Hopkins Plaza Baltimore, Maryland 21201 Telephone: (301)962-2361
		Region IV: Regional Administrator Federal Highway Administration Suite 200 1720 Peachtree Road, N.W. Atlanta, Georgia 30309 Telephone: (404)881-4078

AGENCY	PERSON OR OFFICE TO CONTACT	ADDITIONAL INFORMATION; COMMENTS
Federal Highway Administration (cont)		Region V: Regional Administrator Federal Highway Administration Dixie Highway Homewood, Illinois 60430 Telephone: (312)799-6300 Region VI: Regional Administrator Federal Highway Administration 819 Taylor Street Fort Worth, Texas 76102 Telephone: (817)334-3221 Region VII: Regional Administrator Federal Highway Administration P.O. Box 19715 Kansas City, Missouri 64141 Telephone: (816)926-7565 Region VIII: Regional Administrator Federal Highway Administration Building 40, Room 242 Denver Federal Center Denver, Colorado 80225 Telephone: (303)234-4051 Region IX: Regional Administrator Federal Highway Administration Two Embarcadero Center Suite 530 450 Golden Gate Avenue Box 36096 San Francisco, California 94111 Telephone: (415)556-3951 Region X: Regional Administrator Federal Highway Administration 412 Mohawk Building

AGENCY	PERSON OR OFFICE TO CONTACT	ADDITIONAL INFORMATION; COMMENTS
Federal Highway Adminis- tration (cont)		222 S.W. Morrison Street Portland, Oregon 97204 Telephone: (503)221-2052
Urban Mass Transporta- tion Administra- tion	Office of Program Analysis Urban Mass Transportation Administration 400 7th Street, S.W. Washington, D.C. 20590 Telephone: (202)426-4020	Region I: Office of the Regional Director Urban Mass Transportation Administration Kendall Square Cambridge, Massachusetts 02142 Telephone: (617)494-2055

Region II:
Office of the Regional
Director
Urban Mass Transportation
Administration
Suite 1811
26 Federal Plaza
New York, New York 10007
Telephone: (212)264-8162

Region III:
Office of the Regional
Director
Urban Mass Transportation
Administration
1010 Mall Building
434 Walnut Street
Philadelphia, Pennsylvania
19106
Telephone: (215)597-4179

Region IV:
Office of the Regional
Director
Urban Mass Transportation
Administration
Suite 400
1720 Peachtree Road, N.W.
Atlanta, Georgia 30309
Telephone: (404)881-3948

Region V:
Office of the Regional
Director

AGENCY	PERSON OR OFFICE TO CONTACT	ADDITIONAL INFORMATION; COMMENTS
Urban Mass Transportation Administration (cont)		Urban Mass Transportation Administration Suite 1740 300 South Wacker Drive Chicago, Illinois 60606 Telephone: (312)353-0100 Region VI: Office of the Regional Director Urban Mass Transportation Administration 9A32 Federal Center 819 Taylor Street Fort Worth, Texas 76102 Telephone: (817)334-3787 Region VII: Office of the Regional Director Urban Mass Transportation Administration c/o FAA Management Systems Division Room 303 6301 Rock Hill Road Kansas City, Missouri 64131 Telephone: (816)926-5053 Region VIII: Office of the Regional Director Urban Mass Transportation Administration 1822 Prudential Plaza 1050 17th Street Denver, Colorado 80202 Telephone: (303)837-3242 Region IX: Office of the Regional Director Urban Mass Transportation Administration Suite 620 Two Embarcadero Center

AGENCY	PERSON OR OFFICE TO CONTACT	ADDITIONAL INFORMATION; COMMENTS
Urban Mass Transporta- tion Adminis- tration (cont)		San Francisco, California 94111 Telephone: (415)556-2884 Region X: Office of the Regional Director Urban Mass Transportation Administration 3106 Federal Building 915 Second Avenue Seattle, Washington 98174 Telephone: (206)442-4210
Environmental Protection Agency (EPA)	Director Office of Federal Activities, A-104 Environmental Protection Agency 401 M Street, S.W. Washington, D.C. 20460 Telephone: (202)755-0770	This office should be con- tacted for copies of impact statements on legislation, regu- lations, national program proposals, and major policy issues. For all other EPA con- sultation, contact the regional administrator for the area in which the project is planned. Region I: Regional Administrator U.S. Environmental Protec- tion Agency 2303 John F. Kennedy Fed- eral Building Boston, Massachusetts 02203 Telephone: (617)223-7210 Region II: Regional Administrator U.S. Environmental Protec- tion Agency Room 908 26 Federal Plaza New York, New York 10007 Telephone: (212)264-2525 Region III: Regional Administrator U.S. Environmental Protec- tion Agency

AGENCY	PERSON OR OFFICE TO CONTACT	ADDITIONAL INFORMATION; COMMENTS
Environmental Protection Agency (EPA) (cont)		Curtis Building 6th and Walnut Streets Philadelphia, Pennsylvania 19106 Telephone: (215)597-9801

Region IV:
Regional Administrator
U.S. Environmental Protection Agency
345 Courtland Street, N.E.
Atlanta, Georgia 30308
Telephone: (404)526-5727

Region V:
Regional Administrator
U.S. Environmental Protection Agency
1 N. Wacker Drive
Chicago, Illinois 60606
Telephone: (312)353-5250

Region VI:
Regional Administrator
U.S. Environmental Protection Agency
1st International Building
1201 Elm Street
Dallas, Texas 75270
Telephone: (214)749-1962

Region VII:
Regional Administrator
U.S. Environmental Protection Agency
1735 Baltimore Avenue
Kansas City, Missouri 64108
Telephone: (816)374-5493

Region VIII:
Regional Administrator
U.S. Environmental Protection Agency
900 Lincoln Tower
1860 Lincoln Street
Denver, Colorado 80203
Telephone: (303)837-3895

AGENCY	PERSON OR OFFICE TO CONTACT	ADDITIONAL INFORMATION; COMMENTS
Environmental Protection Agency (EPA) (cont)		Region IX: Regional Administrator U.S. Environmental Protection Agency 100 California Street San Francisco, California 94111 Telephone: (415)556-2320 Region X: Regional Administrator U.S. Environmental Protection Agency 1200 Sixth Avenue Seattle, Washington 98101 Telephone: (206)442-1220
General Services Administration	Mr. Andrew E. Kauders Director of Environmental Affairs Division Special Studies Programs Office General Services Administration 18th and F Streets, N.W. Washington, D.C. 20405 Telephone: (202)566-0405	
Interstate Commerce Commission	Mr. Richard I. Chais, Chief Section of Energy and Environment Interstate Commerce Commission Room 3373 12th and Constitution Avenue, N.W. Washington, D.C. 20423 Telephone: (202)275-7692	
Nuclear Regulatory Commission	Mr. Voss Moore Assistant Director for Environmental Projects Division of Site Safety and Environmental Analysis	Copies of draft impact statements are available from: Ms. Bernadine Scharf Distribution Department

AGENCY	PERSON OR OFFICE TO CONTACT	ADDITIONAL INFORMATION; COMMENTS
Nuclear Regulatory Commission (cont)	Nuclear Regulatory Commission Washington, D.C. 20555 Telephone: (301)492-8446	Nuclear Regulatory Commission Washington, D.C. 20555

Copies of final impact statements may be purchased from:

National Technical Information Service Springfield, Virginia 22161

Copies of final and draft impact statements are available for inspection at:

NRC Public Document Room 1700 H Street, N.W. Washington, D.C.

and at or near the proposed site in the local public document room.

STANDARD FEDERAL REGIONS

223

F

Sources of Further Information on NEPA and the Environmental Impact Statement Process

I. Techniques for Citizen Action

    A Public Citizen's Action Manual by Donald K. Ross. Available free of charge from the Public Interest Research Group, 1346 Connecticut Ave., N.W., Suite 419A, Washington, D.C. 20036. Published by Grossman Publishers, New York, 1973.

    Common Cause Action Manual. Available free of charge from Common Cause, 2030 M St., N.W., Washington, D.C. 20036.

    The End of the Road: A Citizen's Guide to Transportation Problem-Solving. Available from the National Wildlife Federation, 1412 16th St., N.W., Washington, D.C. 20036. Price $3.50. Published by the National Wildlife Federation and Environmental Action Foundation, Inc., 1977.

    The Organizer's Manual by the O.M. Collective. Published by Bantam, New York, 1971.

    The Sierra Club Political Handbook. Available free of charge from the Sierra Club, 530 Bush St., San Francisco, Calif. 94108.

II. Legal Analysis of NEPA's Requirements

    Environmental Law by William H. Rodgers, Jr. Published by West Publishing Company, St. Paul, Minn., 1977, pp. 697–834.

    Federal Environmental Law by Fred Anderson. Edited by Erica L. Dolgin and Thomas G.P. Guilbert. Published by West Publishing Company, St. Paul, Minn., 1974, pp. 238–419.

III. Preparing and Reviewing Impact Statements

    Environmental Impact Assessment by Larry Canter. Published by McGraw-Hill, New York, 1977.

    Guidelines for the Preparation of Environmental Reports for Fossil-fueled Steam Electric Generating Stations. This guide for preparing impact statements on fossil-fueled power plants has been developed by the Department of the Interior. A copy may be obtained by writing to the Office of Environmental Review, Office of the Secretary, Department of the Interior, Washington, D.C. 20240.

    Preparation of Environmental Reports for Nuclear Power Stations. NUREG 0099. This guide for preparing impact statements on nuclear power plants has been developed by the Nuclear Regulatory Commission. A copy may be obtained by writing to the Office of Public Affairs, Nuclear Regulatory Commission, Washington, D.C. 20555.

    The following guides for preparing impact statements have been developed by the Environmental Protection Agency Copies may be obtained free of charge by writing to the Office of Federal Activities, Environmental Protection Agency, 401 M Street, S.W., Washington, D.C. 20460.

Manual for Preparation of EIS's for Waste Water Treatment Works, Facilities Plans, and 208 Areawide Waste Treatment Management Plans

Environmental Impact Assessment Guidelines for Selected New Source Industries

A Guide for Assessing Environmental Impacts. This guide for preparing impact statements on public buildings has been developed by the General Services Administration. A copy may be obtained by writing to The Commissioner, Public Buildings Service, General Services Administration, Washington, D.C. 20405.

Interim Guide for Environmental Assessment: HUD Field Office Edition (1975). This guide for preparing impact statements on housing projects has been developed by the Department of Housing and Urban Development. A copy may be obtained from the Superintendent of Documents, U.S. Government Printing Office, Washington, D.C. 20402.

The following set of guides on preparing impact statements for highway projects has been developed by the Department of Transportation:

The Environmental Assessment Notebook Series: Highways
    Part I: Identification of Transportation Alternatives
    Part II: Social Impacts
    Part III: Economic Impacts
    Part IV: Physical Impacts
    Part V: Organization and Contents of Environmental Assessment Materials
    Part VI: Environmental Assessment Reference Book

The set, numbered GPO #050-000-00109-1, may be obtained from the Superintendent of Documents, U.S. Government Printing Office, Washington, D.C. 20402. Price $21.

The Environmental Assessment of Airport Development Actions (March 1977). This guide for preparing impact statements on airport projects has been developed by the Federal Aviation Administration. A copy may be obtained by writing to the Office of Airport Programs, Federal Aviation Administration, 800 Independence Ave., S.W., Washington, D.C. 20591.

The following guides on reviewing impact statements have been developed by the Environmental Protection Agency. Copies may be obtained free of charge by writing to the Office of Federal Activities, Environmental Protection Agency, 401 M Street, S.W., Washington, D.C. 20460.

Guidelines for the Review of Environmental Impact Statements
  Volume I: Highway Projects
  Volume II: Airports (not completed as of April 1, 1978)
  Volume III: Impoundment Projects
  Volume IV: Channelization Projects

Guidelines for the Review of Environmental Impact Statements for Nuclear Power Plants (Light Water Cooled Reactors)
  Part I–Procedures
  Part II–Technical Analysis

Guidelines for the Environmental Review of Power Plant Cooling Lakes

Best Practices for New Surface and Underground Coal Mines

IV.   Studies of Agency Implementation of NEPA

Environmental Impact Statements: An Analysis of Six Years' Experience by Seventy Federal Agencies by the Council on Environment Quality, 1976. Available free of charge from the Council on Environmental Quality, 722 Jackson Pl., N.W., Washington, D.C. 20006. Published by the U.S. Government Printing Office, Washington, D.C. 20402.

Environmental Impact Statements–A Report of the Commission on Federal Paperwork by the Commission on Federal Paperwork, 1977. Available from the U.S. Government Printing Office, Washington, D.C. 20402. GPO #040-000-00382-7.

Environmental Policy and Administrative Change by Richard N.L. Andrews. Published by Lexington Books, Lexington, Mass., 1976.

Natural Resources Journal—Special issue developed for a symposium on environmental impact statements. Volume 16, Number 2, April 1976. Available from the University of New Mexico School of Law, Albuquerque, New Mexico 87131. Price $4.25.

The Environmental Impact Statement—It Seldom Causes Long Project Delays But Could Be More Useful If Prepared Earlier. Available free of charge from the U.S. General Accounting Office, 441 G St., N.W., Washington, D.C. 20548.

A National Policy for the Environment: NEPA and Its Aftermath by Richard Liroff. Published by Indiana University Press, 1976.

The Environmental Impact Statement vs. The Real World by

Eugene Bardach and Lucian Pugliaresi. Published in *The Public Interest*, 49:545–553, November 1977.

V.   New Developments Under the Statute

Environment Reporter. Published by the Bureau of National Affairs, Washington, D.C. Issued weekly. Available in large law libraries.

Environmental Law Reporter. Published by the Environmental Law Institute, Washington, D.C. Issued monthly. Available in large law libraries.

Environmental Quality, Annual Report of the Council on Environmental Quality (1977 Annual Report). Available free of charge from the Council on Environmental Quality, 722 Jackson Pl., N.W., Washington, D.C. 20006. Issued annually. Published by the U.S. Government Printing Office, Washington, D.C. 20402. GPO #041-011-00035-1.

The 102 Monitor by U.S. Environmental Protection Agency. Available from the U.S. Government Printing Office, Washington, D.C. 20402. Issued monthly. Annual subscription: $21.50.

G

State Environmental Impact Statement Requirements

Twenty-nine states and Puerto Rico have established an environmental impact statement process through legislation, executive order, or administrative agency directive. Each requires an analysis of the environmental effects of projects, provides an opportunity for the public to comment on the analysis, and directs that the analysis and comments be considered by agencies in their decision making. Although a few of the states do not refer to their requirements as an "impact statement process," all of them follow the general pattern established by NEPA, through which the public is involved in the review of projects that significantly affect the environment.

The listing on pages 235-242 lists these state requirements and divides them into two general categories: "comprehensive requirements" and "limited requirements." A comprehensive requirement is one that applies to a wide variety of projects in the state. A limited requirement is one that applies to only a few categories of projects in the state—for example, only the construction of electric power plants or only actions affecting the state's coastal zone. As of April 1, 1978, 17 states and Puerto Rico had comprehensive impact statement requirements; 12 states had limited impact statement requirements.

Applicability of the Requirement. California, Massachusetts, New York, Washington, and Puerto Rico have impact statement provisions with the broadest applicability; they require statements on all projects of state and local agencies that will significantly affect the environment. They also require statements for privately sponsored projects that will significantly affect the environment and that involve either public funds or a state or local permit. Minnesota is unique in requiring an impact statement for any major private project of more than local significance, even though no government funds or permits are associated with the project.

A number of states, such as Maryland and Virginia, restrict their impact statement requirements to actions of state agencies. Under New Jersey's Executive Order No. 53, impact statements need only be prepared for actions that have values greater than one million dollars, or for actions that would affect areas which the state determines are "critical," such as flood plains. South Carolina restricts its environmental impact statement requirement to actions affecting coastal tidelines and wetlands. Florida requires impact statements for actions with regional impact or which affect critical areas. In Georgia, environmental impact statements are required only for state-funded toll roads. In Arkansas, Nevada, and New Hampshire, statements are required only in connection with the siting of public utilities and transmission lines. South Dakota is the only state that explicitly exempts agencies involved in environmental protection activities from the requirement to write impact statements.

Content of the Impact Statement. Some jurisdictions, such as Washington, Indiana, and Puerto Rico, follow the wording in Section 102(2)(c) of NEPA on the topics that must be covered in an impact statement. Most states with comprehensive requirements, however, have supplemented NEPA's provisions. For example, Maryland, Massachusetts, California, and New

York also require a description of mitigation measures: measures that will be taken to minimize or reduce the environmental damage, disclosed in the impact statement, that will result from the proposed project. New York, California, and South Dakota require a discussion of the project's growth-inducing aspects. Connecticut, New York, and California require a description of the project's effect on energy conservation and consumption. The costs and benefits of the project must be described under Connecticut's law; and Wisconsin requires a discussion of the short- and long-term beneficial aspects of the project, as well as its economic advantages and disadvantages. Impact statements prepared in Virginia and New Jersey (under the Coastal Zone Law) must set forth reasons for the rejection of the alternatives to the proposed project. Minnesota requires a description of the impact on state government of federal controls associated with the project.

Most states with "limited" impact statement requirements, instead of following the general provisions of Section 102(2)(c) of NEPA, list more specific topics that must be covered. For example, Arkansas's "Utility Facility Environmental and Economic Protection Act" requires a description of the project's effect on land, air, and water environments, on established park and recreational areas, and on sites of natural, scenic, and historic resources. In approving utility sites, the New Hampshire Public Utilities Commission must find that the site will not unduly interfere with the orderly development of the region and will not have an unreasonable adverse effect on aesthetics, historic sites, air and water quality, the natural environment, and public health and safety. The Arizona Power Plant and Transmission Line Siting Committee must consider in its impact statement other existing development plans; fish, wildlife, and plant life; noise emissions; public recreation; scenic areas and historical and archeological sites; unique biological habitats; technical practicability of achieving the proposed objective; and estimated facility cost.

Provisions for Involvement of the Public. Although not all states actively solicit public comments on impact statements, all the states provide an opportunity for individuals to express their views on a project and require that those views be considered in the agency's decision making. New York's statute further requires that agencies respond to all the comments submitted. New Jersey's Executive Order No. 53 constrains public involvement by allowing only a four-week period for comments on a statement.

The decision on whether a public hearing should be held generally is within the full discretion of the agency, although in some states there are limits on the exercise of this discretion. In New York, in reaching a decision on whether to hold a hearing, agencies are required to consider the degree of interest in the project and the extent to which a hearing would aid the decision-making process by providing a forum for the collection of public comments. In Nevada, if there is no public protest to the notice of the proposed construction of a utility's facility, the Public

Service Commission may dispense with its normal hearing requirement. Under Connecticut law, a public hearing must be held if 25 or more people request it. Some states, such as Utah and Michigan, simply rely upon general administrative procedures and agency regulations in determining whether a public hearing should be held. A few states, such as Arkansas (utilities), Delaware (projects affecting critical areas), and New Hampshire (utilities), permit no discretion on the part of an agency; they direct that public hearings be held in all cases.

Implementation of the Requirements. In Minnesota, the Minnesota Environmental Quality Board (a board composed of department heads and citizens) oversees the state's impact statement process. It also is empowered to require revision of an impact statement which it deems inadequate. Other states, such as South Dakota, have no legislatively designated agency to oversee the process. Hawaii and Virginia require that the governor review and accept impact statements; Hawaii also provides for review and acceptance by mayors. In Michigan, an interdepartmental review committee conducts an initial technical review of each impact statement. California and New York do not empower any agency to oversee the process, but they do require that statements be circulated to other agencies in the state for their comments.

The number of impact statements prepared in each state varies greatly. From 1970–1975, California prepared approximately 3,800 statements. Washington and Wisconsin prepare approximately 50 statements per year. Hawaii prepared 35 statements in 1976 and 45 statements in 1977. In Connecticut, only six statements have been prepared since 1975.

While the number of statements written each year provides an indication of the importance of the state's EIS process, the figures should be used with some caution. A small number of statements does not necessarily mean that few projects are scrutinized for their possible environmental impact. For example, Montana prepared 10 impact statements in 1976, and 13 statements in 1977; however, it prepared 193 preliminary environmental reports in 1976 (which formed the basis for deciding whether to prepare a full environmental impact statement) and 216 preliminary environmental reports in 1977.

A few state laws, unlike NEPA, require an agency to make written findings in support of its decision. In New York, agencies must make a formal determination that, from the alternatives available, the one chosen will, consistent with social, economic, and other essential considerations, result in the minimum adverse environmental impact. In South Dakota, agencies must find that all feasible action will be taken to minimize or avoid environmental problems revealed by the impact statement process. The Nevada Public Service Commission must find that, considering technical and economic factors, the utility's proposed facility represents the minimum adverse environmental impact. With a few other exceptions, such as California and New Hampshire, the remaining states do not explicitly require formal findings.

STATE ENVIRONMENTAL IMPACT STATEMENT REQUIREMENTS
April 1, 1978
States with Comprehensive Statutory Requirements

California

Source: California Environmental Quality Act, Cal. Pub. Res. Code Section 21000 *et seq.* (Cum. Supp. 1978) (West).

Guidelines: *Guidelines for Implementation of the California Environmental Quality Act of 1970,* Cal. Admin. Code tit. 14, Section 15000 *et seq.,* as amended March 4, 1978. Guidelines are issued by the Resources Agency of California.

Contact: Norman E. Hill, Assistant to the Secretary for Resources, The Resources Agency, 1416 Ninth Street, Sacramento, California 95814 (Telephone: 916/445-9134).

Connecticut

Source: Connecticut Environmental Policy Act of 1973, Conn. Gen. Stat. Ann. Section 22a-1a *et seq.* (Cum. Supp. 1978) (West).

Guidelines: No guidelines have been issued.

Contact: Jonathan Clapp, Senior Environmental Analyst, Planning and Coordination Unit, Department of Environmental Protection, 118 State Office Building, Hartford, Connecticut 06115 (Telephone: 203/566-3740).

Hawaii

Source: Governor's Executive Order of August 21, 1974, as supplemented by Environmental Quality Commission and Environmental Impact Statements, Haw. Rev. Stat. Section 343-1 *et seq.* (1976 Replacement).

Guidelines: *Rules of Practice and Procedure* and *Environmental Impact Statement Requirements,* issued by the Environmental Quality Commission on June 2, 1975.

Contact: Donald A. Dremner, Chairman, Environmental Quality Commission, Office of the Governor, Room 301, 550 Halekauwila Street, Honolulu, Hawaii 96813 (Telephone: 808/548-6915).

Indiana

Source: Indiana Environmental Policy Act, Ind. Code Ann. Section 13-1-10-1 *et seq.* (1973) (Burns).

Guidelines: *EMB-2 and EMB-3,* Ind. Admin. Rules and Regs. Ann. Section 13-1-10-3 *et seq.* and Section 13-7-5-1 *et seq.* (Supp. 1977). The Environmental Management Board prepares guidelines.

Contact: L. Robert Carter, Coordinator of Environmental Programs, Indiana State Board of Health, 1330 West Michigan Street, Indianapolis, Indiana 46206 (Telephone: 317/633-8467).

Maryland

Source: Maryland Environmental Policy Act of 1973, Md. Nat. Res. Code Ann. Section 1-301 *et seq.* (Cum. Supp. 1977).

Guidelines: *Revised Guidelines for Implementation of the Maryland Environ-mental Policy Act,* issued by the Secretary of the Department of Natural Resources, June 15, 1974.

Contact: Joseph Knapp, Administrator, Clearing House Review, Department of Natural Resources, Tawes State Office Building, Annapolis, Maryland 21401 (Telephone: 301/369-3548).

Massachusetts

Source: Massachusetts Environmental Policy Act, Mass. Ann. Laws Ch. 30, Sections 61 and 62 (Cum. Supp. 1977) (Michie/Law. Co-op), as amended by 1978 Mass. Acts Ch. 947, January 10, 1978.

Guidelines: *Regulations Governing the Implementation of the Massachusetts Environmental Policy Act,* issued by the Executive Office of Environ-mental Affairs on February 16, 1978.

Contact: William Hicks, Director, Massachusetts Environmental Impact Re-view Office, Executive Office of Environmental Affairs, Room 2001, 100 Cambridge Street, Boston, Massachusetts 02202 (Telephone: 617/727-5830).

Minnesota

Source: Minnesota Environmental Policy Act of 1973, Minn. Stat. Ann. Section 116D.01 *et seq.* (1977) (West).

Guidelines: *Rules and Regulations for Environmental Impact Statements,* issued by the Minnesota Environmental Quality Council on April 4, 1974, and amended on February 13, 1977.

Contact: Joe Sizer, Director, Environmental Planning, Environmental Quality Board, Capital Square Building, 550 Cedar Street, St. Paul, Minnesota 55101 (Telephone: 612/296-2712).

Montana

Source: Montana Environmental Policy Act, Mont. Rev. Codes Ann. Section 69-6501 *et seq.* (Cum. Supp. 1977).

Guidelines: *Uniform Rules Implementing the Montana Environmental Policy Act.* Final version adopted by the Montana Commission on Environmental Quality on January 15, 1976.

Contact: Eileen Shore, Staff Attorney, Montana Commission on Environmental Quality, Capitol Station, Helena, Montana 59601 (Telephone: 406/449-3742).

New York

Source: New York State Environmental Quality Review Act, N.Y. Envir. Conserv. Law Section 8-0101 *et seq.* (Cum. Supp. 1977-1978) (McKinney) as amended by 1976 N.Y. Laws, Ch. 228, Section 5 and 1977 N.Y. Laws, Ch. 252, Section 9, *et seq.*

Guidelines: 6 N.Y.C.R.R. Park 617, revised January 24, 1978 by the Department of Environmental Conservation.

Contact: Allen Davis, Environmental Quality Review Section, Office of Envi-ronmental Analysis, New York State Department of Environmental

Conservation, 50 Wolf Road, Albany, New York 12233 (Telephone: 518/457-2224).

North Carolina

Source: North Carolina Environmental Policy Act of 1971, N.C. Gen. Stat. Section 113A-1 (1975 Replacement).

Guidelines: North Carolina Department of Administration, *Guidelines for the Implementation of the Environmental Policy Act of 1971*, revised March 1, 1975.

Contact: Anne Taylor, Policy Advisor for Natural Resources, Division of Policy Development, Department of Administration, 116 West Jones Street, Raleigh, North Carolina 27603 (Telephone: 919/733-4131).

Puerto Rico

Source: Public Environmental Policy Act, P.R. Laws Ann. tit. 12, Section 1121 *et seq.* (Cum. Supp. 1976).

Guidelines: *Guidelines for the Preparation, Evaluation, and Use of Environmental Impact Statements*, issued by the Environmental Quality Board on December 19, 1972.

Contact: Roberto Rexach, Executive Director, Environmental Quality Board, 4th Floor, 1550 Ponce de Leon Avenue, Santurce, Puerto Rico 19910 (Telephone: 809/725-5140).

South Dakota

Source: South Dakota Environmental Policy Act, S.D. Codified Laws Section 34A-9-1 *et seq.* (1977 Revision).

Guidelines: Informal guidelines issued by the Department of Environmental Protection in 1974.

Contact: Harold Lenhart, Deputy Secretary, South Dakota Department of Environmental Protection, Foss Building, Pierre, South Dakota 57501 (Telephone: 605/224-3351).

Virginia

Source: Virginia Environmental Quality Act of 1973, Va. Code Section 10-17.107 *et seq.* (Cum. Supp. 1977).

Guidelines: *Procedures Manual for Environmental Impact Statements in the Commonwealth of Virginia*, revised June 1976, by the Governor's Council on the Environment.

Contact: Reginald Wallace, Environmental Impact Statement Coordinator, Council on the Environment, 903 9th Street Office Building, Richmond, Virginia 23219 (Telephone: 804/786-4500).

Washington

Source: State Environmental Policy Act, Wash. Rev. Code Ann. Section 43.21C.010 *et seq.* (Supp. 1976).

Guidelines: *State Environmental Policy Act Guidelines*, Wash. Admin. Code Section 197-10, revised January 21, 1978, by the Department of Ecology.

Contact: Peter R. Haskin, Environmental Review and Evaluation, Office of
                Planning and Program Development, Department of Ecology,
                Olympia, Washington 98504 (Telephone: 206/753-6890).

Wisconsin

Source: Wisconsin Environmental Policy Act of 1971, Wis. Stat. Ann. Section
                1.11 (Cum. Supp. 1977-1978) (West).
Guidelines: *Revised Guidelines for the Implementation of the Wisconsin Envi-*
                ronmental Policy Act, issued by Governor's Executive Order No. 26
                (February, 1976).
Contact: Caryl Terrell, State WEPA Coordinator, Office of State Planning and
                Energy, Room B-130, 1 West Wilson Street, Madison, Wisconsin 53702
                (Telephone: 608/266-1718).

States with Comprehensive Executive or Administrative Orders

Michigan

Source: Michigan Executive Order 1974-4 (May, 1974).
Guidelines: *Guidelines for the Preparation and Review of Environmental Impact*
                Statements under Executive Order 1974-4, issued by the Environ-
                mental Review Board in November, 1975.
Contact: Terry L. Yonker, Executive Secretary, Environmental Review Board,
                Department of Management and Budget, Lansing, Michigan 48913
                (Telephone: 517/373-6491).

New Jersey

Source: New Jersey Executive Order No. 53 (October 15, 1973).
Guidelines: *Guidelines for the Preparation of an Environmental Impact State-*
                ment, issued by the Office of the Commissioner, Department of
                Environmental Protection, and revised in February 1974.
Contact: Lawrence Schmidt, Chief, Office of Environmental Review, Depart-
                ment of Environmental Protection, P. O. Box 1390, Trenton, New
                Jersey 08625 (Telephone: 609/292-2662).

Utah

Source: State of Utah Executive Order, August 27, 1974.
Guidelines: No guidelines have been issued.
Contact: William C. Quigley, Assistant Attorney General, Room 236, Office
                of the Attorney General, State Capitol Building, Salt Lake City, Utah
                84114 (Telephone: 801/533-7643).

States with Limited EIS Requirements

Arizona

Source: (a) Game and Fish Commission Policy of July 2, 1971.
                (b) Power Plant Transmission Line Siting Act, Ariz. Rev. Stat. Section
                    40-360 *et seq.* (Cum. Supp. 1977-1978).

Guidelines: (a) Memorandum by the Arizona Game and Fish Commission, "Requirements for Environmental Impact Statements," issued June 9, 1971.

(b) No guidelines have been issued.

Contact: (a) Robert D. Curtis, Chief, Wildlife Planning and Development Division, Arizona Game and Fish Department, 2222 West Greenway Road, Phoenix, Arizona 85023 (Telephone: 602/942-3000).

(b) Chairman, Power Plant Transmission Line Siting Committee, Attorney General's Office, Room 200, 1700 West Washington, Phoenix, Arizona 85007 (Telephone: 602/271-4266).

Arkansas

Source: Utility Facility Environmental and Economic Protection Act, Ark. Stat. Ann. Section 73-276 *et seq.* (Cum. Supp. 1977).

Guidelines: Informal guidelines issued by the Arkansas Public Service Commission.

Contact: Stephen Cuffman, Counsel, Public Service Commission, Justice Building, Little Rock, Arkansas 72202 (Telephone: 501/371-2040).

Delaware

Source: (a) Delaware Coastal Zone Act, Del. Code Ann. tit. 7, Section 7001 *et seq.* (1974).

(b) The Wetlands Act, Del. Code Ann. tit. 7, Section 6601 *et seq.* (1974).

Guidelines: (a) *Permit Application Instructions and Forms and Information Material on Required Procedures for the Coastal Zone Act*, adopted by the Delaware Office of Management, Budget, and Planning on July 1, 1977.

(b) *Wetlands Regulations*, adopted by the Department of Natural Resources and Environmental Control, December 23, 1976.

Contact: (a) David Hugg, Manager, Coastal Management Program, Delaware Office of Management, Budget, and Planning, Dover, Delaware 19901 (Telephone: 302/678-4271).

(b) William Moyer, Wetlands Manager, Department of Natural Resources and Environmental Control, Division of Environmental Control, Dover, Delaware 19901 (Telephone: 302/678-4761).

Florida

Source: The Florida Environmental Land and Water Management Act of 1972, Fla. Stat. Ann. Section 380.012 *et seq.* (Cum. Supp. 1977) (West).

Guidelines: *Developments Presumed to be of Regional Impact*, Fla. Admin. Code Ch. 22F-2 (1978).

Contact: James May, Bureau Chief, Bureau of Land and Water Management, Division of State Planning, Department of Administration, 660 Apalachee Parkway, Tallahassee, Florida 32304 (Telephone: 904/488-4925).

Georgia

Source: State Tollway Authority Act, Ga. Code Ann. Section 95a-1241(e)(1) (1976).
Guidelines: *Policy and Procedures Manual: State Tollway Authority*, revised by Georgia's Tollway Administrator's Office in February 1973.
Contact: Robert L. Austin, State Location Engineer, Division of Preconstruction, Department of Transportation, 2 Capitol Square, Atlanta, Georgia 30334 (Telephone: 404/656-5312).

Maine

Source: Site Location Law, Me. Rev. Stat. tit. 38, Section 481 *et seq.* (Cum. Supp. 1977-1978).
Guidelines: *Site Law Regulations and Guidelines*, issued by the Department of Environmental Protection.
Contact: Information and Education Division, Department of Environmental Protection, State House, Augusta, Maine 04333 (Telephone: 207/289-2691).

Mississippi

Source: Coastal Wetlands Protection Law, Miss. Code Ann. Section 49-27-1 *et seq.* (Cum. Supp. 1977), to be amended by Sen. Bill 3498 (1978) (passed but not yet signed as of April 1, 1978).
Guidelines: *Rules and Regulations Pertaining to the Coastal Wetlands Protection Law*, revised by Mississippi Marine Resources Council, April 15, 1975.
Contact: Joe Gill, Jr., Marine Projects Manager, Mississippi Marine Resources Council, P. O. Drawer 959, Long Beach, Mississippi 39560 (Telephone: 601/864-4602).

Nebraska

Source and   Nebraska Department of Roads, Department of Roads Action Plan
Guidelines: (1973), as revised by the State of Nebraska Environmental Action Plan, prepared by the Nebraska Department of Roads and approved by the Federal Highway Administration, June 24, 1975.
Contact: Robert O. Kuzelka, Comprehensive Planning Coordinator, Office of Planning and Programming, P. O. Box 94601, State Capital, Lincoln, Nebraska 68509 (Telephone: 402/471-2414).

Nevada

Source: Utility Environmental Protection Act, Nev. Rev. Stat. Section 704.820 *et seq.* (1973).
Guidelines: No guidelines have been issued.
Contact: Heber Hardy, Chairman, Public Service Commission of Nevada, Kinkead Building, Carson City, Nevada 89701 (Telephone: 702/885-4180).

New Hampshire

Source: Electric Power Plant Transmission Line Siting and Construction Procedure, N.H. Rev. Stat. Ann. Section 162-F:1 *et seq.* (Supp. 1975).

Guidelines: No guidelines have been issued.
Contact: Dom D'Ambruoso, Secretary of Public Utilities Commission, 8 Old
              Suncook Road, Concord, New Hampshire 03301 (Telephone: 603/
              271-2443).

New Jersey

Source: (a) Coastal Area Facility Review Act, N.J. Stat. Ann. Section
                  13:19-1 *et seq.* (Cum. Supp. 1977-1978).
              (b) The New Jersey Wetlands Act of 1970, N.J. Stat. Ann. Section
                  13:9A-1 *et seq.* (Cum. Supp. 1977-1978).
Guidelines: (a) *CAFRA Rules and Regulations*, N.J.A.C. 7:7D-1.1 *et seq.*, effective
                  November 18, 1975.
              (b) *Procedural Rules and Regulations*, N.J.A.C. 7:7A-1.1 *et seq.*,
                  revised September 3, 1976.
Contact: (a) David N. Kinsey, Chief, Office of Coastal Zone Management,
                  New Jersey Department of Environmental Protection, P. O. Box
                  1889, Trenton, New Jersey 08625 (Telephone: 609/292-8262).
              (b) Thomas F. Hampton, Supervisor, Office of Wetlands Manage-
                  ment, Division of Marine Services, Department of Environmental
                  Protection, P. O. Box 1889, Trenton, New Jersey 08625 (Tele-
                  phone: 609/292-8202).

South Carolina

Source: South Carolina Coastal Management Act, S.C. Code Section 48-39-10
              et seq. (Cum. Supp. 1977).
Guidelines: *Final Rules and Regulations for Permits of Alterations of Critical
              Areas*, issued by the South Carolina Coastal Council (1977) in the
              South Carolina State Register.
Contact: Wayne Beam, Executive Director, South Carolina Coastal Council,
              116 Bankers Trust Tower, Columbia, South Carolina 29201 (Tele-
              phone: 803/758-8442).

City EIS Requirements

Bowie, Maryland

Source and   The Bowie, Maryland Environmental Policy and Impact Statement
Guidelines: Ordinance, passed by the City Council on May 3, 1971; Ordinance
              0-2-73, Declaring an Environmental Policy and Providing for Environ-
              mental Impact Statements, passed by the City Council on July 16,
              1973; and Ordinance 0-14-76, Changing Notification and Referral
              Requirements under the Ordinance, passed by the City Council on
              September 8, 1976.
Contact: Bruce Allen, Acting Planning Director, Office of Planning and Com-
              munity Development, City Hall, Bowie, Maryland 20715 (Telephone:
              301/262-6200).

New York City

Source and   Executive Order No. 87, October 1973 and Executive Order No. 91,
Guidelines: June 1, 1977.
Contact: Dorothy Green, Director, Office of Environmental Impact, New York
                 City Department of Environmental Protection, 2344 Municipal Build-
                 ing, New York, New York 10007 (Telephone: 212/556-4107).

A-11 79 1817 Q

The Author

Neil Orloff is an Associate Professor at Cornell University, where he is on the faculty of the Program on Science, Technology, and Society and the College of Engineering and directs the Cornell Project on Environmental Impact Statements. Professor Orloff holds degrees in law, business, and engineering from Columbia Law School, Harvard Business School, and the Massachusetts Institute of Technology. Prior to assuming his present post, he served as a senior staff member with the Environmental Protection Agency and as Legal Counsel to the President's Council on Environmental Quality, in which capacity he helped prepare the CEQ guidelines for the preparation of environmental impact statements. Professor Orloff is the author of the forthcoming "Casebook on the Legal Aspects of the National Environmental Policy Act."

Date Due